Modeling and Advanced Control for Process Industries

Applications to Paper Making Processes

Ming Rao, Qijun Xia and Yiqun Ying

Modeling and Advanced Control for Process Industries

Applications to Paper Making Processes

With 115 Figures

Springer-Verlag
London Berlin Heidelberg New York
Paris Tokyo Hong Kong
Barcelona Budapest

Ming Rao, PhD
Qijun Xia, PhD, MSc, BSc
Yiqun Ying, MSc, BSc

Department of Chemical Engineering
University of Alberta
536 Chemical-Mineral Engineering Building
Edmonton, Canada T6G 2G6

ISBN 3–540–19881–4 Springer-Verlag Berlin Heidelberg New York
ISBN 0–387–19881–4 Springer-Verlag New York Berlin Heidelberg

British Library Cataloguing in Publication Data
A catalogue record for this book is available from the British Library

Library of Congress Cataloging-in-Publication Data
A catalog record for this book is available from the Library of Congress

Typesetting: Camera-ready by authors
Printed by Athenæum Press Ltd, Newcastle upon Tyne
69/3830–543210 Printed on acid-free paper

PREFACE

The paper machine is a very important part of pulp and paper manufacturing processes. Process modeling and control play key roles in the paper machine operation. Due to the complexity of the process operation and the requirements of high quality product, low cost production, safety and environment protection, more and more pulp and paper companies are looking for advanced control technology to improve their process operation.

This book reports our research results on the modeling and advanced control for paper machines. Both theoretic fundamentals and industrial applications are presented. This is a book in which all the advanced technologies in modeling and control discussed are focused on applications to paper machines. The book is organized as follows: Chapter 1 gives a brief introduction to paper making process fundamentals and an overview of paper machine control. Various process dynamics analysis and modeling techniques are discussed in detail in Chapter 2. Based on the characteristics of paper machine operation, some typical advanced control strategies, such as robust control (Chapter 3), predictive control (Chapter 4), bilinear control (Chapter 5), fault-tolerant control (Chapter 6) and their design and implementation techniques as well as real industrial applications are presented. Since model-based control systems cannot handle the ill-formulated problems involved in paper machines, fuzzy control (Chapter 7), expert systems (Chapter 8), artificial neural networks (Chapter 9) and intelligent on-line monitoring and control systems (Chapter 10) are then introduced. Their applications to process control, control system design, process modeling and product quality prediction, and on-line monitoring and control for paper and pulp processes are also presented. It should be pointed out that not all the control technologies applied to paper machines have been covered in this book. Some important applications, such as cross-machine direction (CD) control, are not included.

This book is designed as a reference book for engineers and research scientists who work on process control, especially in the pulp and paper industry. It can also be used as a textbook for graduate student courses on process modeling and advanced process control in universities.

We hope that this book will help to narrow the gap between academic research and industry application, and to reduce the barriers that exist in applying advanced control technologies to real industrial processes.

SERIES EDITOR'S FOREWORD

The series *Advances in Industrial Control* aims to report and encourage technology transfer in control engineering. The rapid development of control technology impacts all areas of the control discipline. New theory, controllers, actuators, sensors, new industrial processes, computing methods, applications, philosophies, . . . new challenges. Much of this development work resides in industrial reports, feasibility study papers and the reports of advanced collaborative projects. The series offers an opportunity for researchers to present an extended exposition of such new work in all aspects of industrial control for wider and rapid dissemination.

It is always valuable to welcome to the Series a text which deals with the control problems of one individual sector. Dr Rao and his colleagues have developed an Intelligent Online Monitoring and Control System (IOMCS) for pulp and paper making process control applications. In this volume, the background research and the development path for the IOMCS is presented. It is a route which takes in process model developments, advanced control methods and methods like fuzzy control before arriving at the integrated construction of the IOMCS. The whole development is underpinned by the technological framework of a plant-wide distributed control system. For this reason, the concept of an IOMCS has a wider applicability to general process control engineering. We hope the volume will be read with interest by this broader audience of control, manufacturing, process and even production engineers.

M. J. Grimble and M. A. Johnson
Industrial Control Centre,
University of Strathclyde,
Scotland, U.K.

Professor M.J. Grimble and Dr M.A. Johnson, editors of the *Advances in Industrial Control Series*, provided us with very important suggestions and assistance in preparing this book.

Professor Y. Sun and Professor C. Zhou of Zhejiang University gave us the valuable suggestions, allowed us to use the research results from their Institute. We gratefully acknowledge Dr P. Li and Dr Q.G. Wang for their important contributions to Chapter 3 and Chapter 7. Professor H. Qiu, Professor L. Peng and Dr X. Shen helped us by reviewing the manuscript, and providing many useful comments for improvement of the book. Graduate students, H. Fazadeh, J. Sun and J. Zurcher and Dr Q. Wang provided technical support to the content of this book. We would like to express our appreciation for their help and contributions.

We also gratefully acknowledge the financial and technical support of the Natural Sciences and Engineering Research Council of Canada, Canadian Pulp and Paper Association, Weyerhaeuser Canada Grande Prairie Operations, Slave Lake Pulp Corporation, DMI Peace River Pulp Division, MoDo Chemetics, Perde Enterprise, and Canada Alberta Partnership on Forestry.

Ming Rao
Qijun Xia
Yiqun Ying
Edmonton, Canada
October 31, 1993

TABLE OF CONTENTS

CHAPTER 1
BACKGROUND

It is expected that most readers will have some knowledge of the paper-making process and its control systems. However, for those who do not, this chapter is offered as a very brief introduction to the process operation and its control systems, to provide a necessary background about the modeling and control techniques described in this book.

1.1 Paper-making: Process Fundamentals

Paper is a structure formed mainly from natural vegetable fibres, principally obtained from wood, with or without various additives. By selecting the types of fibres, additives and their treatment in the process, a very wide range of pulp and paper products is made.

By pouring the fibre suspension on to an endless, moving sieve-like belt, a mat of fibres is formed continuously. When sufficient water has been removed, this mat is separated from the endless belt and forms a continuous web of paper. The formation of a continuous web from a dilute suspension is the focal point of the paper-making process. The typical paper making process can be broadly divided into the following operations:

Pulping, the extraction of the raw fibrous material from its source.

Stock preparation, the treatment and proportioning of the fibres and additives in the fluid state.

Paper-making, which takes the prepared stock, carries out the final operations prior to the formation of the sheet, followed by pressing and drying, calendering and reeling.

After-processes, e.g. rewinding, coating, and super-calendering.

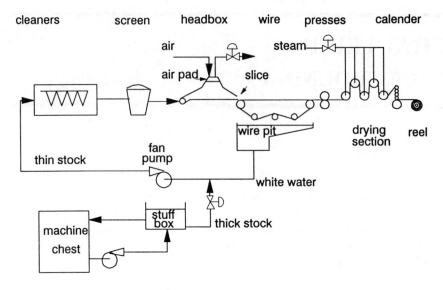

Figure 1.1: Simplified diagram of Fourdrinier paper machine

Finishing end, where the paper reels are converted into cut sheets, sorted, counted, wrapped, stored and despatched. If paper is sold on the reel, the finishing end's function is limited to sampling and checking, then wrapping and despatching.

The process varies from mill to mill and even within the same mill, differences occur according to the product requirements.

Almost all paper is produced continuously. There are various types of paper machines which perform this operation. The Fourdrinier paper machine is one of the most commonly used paper machines (McGill, 1980). Figure 1.1 shows a basic Fourdrinier machine, which has been greatly simplified. The forming part consists of an endless belt of open mesh, called wire, which is driven at the required speed. The diluted stock is poured on at one end, after drainage of most of the water, the sheet formed on the wire is strong enough to be peeled off the wire and transported to carrying felts to the next operation stage, i.e. pressing, which is followed by drying, calendering and reeling.

A typical Fourdrinier paper machine consists of headbox, wire and press, drying section, calenders, drive and finishing part. We discuss these parts separately as follows:

Headbox

At the upstream end of the wire, the suspension, in the form of a thin strip covering the full width of the wire, is poured on to the wire. The process element is called the headbox or flowbox.

The conversion of the pipe flow, which the headbox accepts, into an evenly-distributed wide ribbon flow, is one of the most difficult and most critical operations in the paper-making process. The emergent stream is controlled by the 'slice', a kind of gate valve, which can be manipulated to produce the flow configuration.

One of the most important variables controlled by the headbox and its slice is the velocity at which the stuff emerges on to the wire. This velocity is a function of the pressure behind the slice. Headbox may be open or closed at the top. In open boxes, the stream velocity depends on the level of stock behind the slice. Closed boxes may be pressurized, in which case a fixed level may be chosen, and the total head is the sum of the static head and the pressure above stock level. The emergent velocity will affect formation, the pattern which the fibres take up relative to each other. This in turn has the critical effect on the characteristics of the sheet. In practice, the velocity of stuff flow at the slice is usually fairly close to the velocity of the wire. Another main variable is flow rate, which is related to the consistency of the stock at this point and which also has a critical effect on formation.

Wire and Press

The wire itself is a woven fabrication (like gauze). In the past, the wire was woven from metal. Plastic woven fabrics are now becoming common. The wire is stretched between rolls, the main ones being the breast roll at the slice end and the couch at the take-off end. The table part of the wire where the sheet is forming is supported by small rolls (called table rolls) or by foils. These rolls not only support the wire, but also play an important role in the drainage of water.

To improve paper formation, oscillation is introduced in a cross-machine direction at the wire part at the slice end. This is called shake. At the certain points along the wire, the rate of drainage by gravity, assisted by foils or table rolls, becomes very small. More water is removed by providing a vacuum under the wire. The components which perform this operation are known as suction boxes or vacuum boxes. They are connected to a vacuum pump system. The couch roll at the take-off point may be a plain roll or be provided with a vacuum arrangement.

The underside of the wire is carried by the return rolls, which usually includes at least one that may be varied in angle relative to the wire to guide it, i.e. to keep it running in a central position.

The water which is drained from the sheet is called backwater. It is collected via the wire trays and send to recycle vessels, e.g. the backwell or backwater silo from which most of it is pumped to dilute the incoming stock. Excess backwater may be stored to fill beaters or pumps etc., thus retaining more fines and loading.

Under the wire is the hog pit where the web is directed at start-ups and breaks. The hog pit also commonly receives the deckle trim (deckle being the name for the cross-machine width of the sheet) during running. Some machines have deckle traps which are positioned to provide the required width. On other machines, almost the full width of the wire is used at all times. The required deckle width is separated out just before the couch by fine water sprays, the trim dropping into the hog pit. This pit is fitted with an agitating device, which repulps the wet sheet and returns the pulp to one of the storage chests.

The sheet, on leaving the wire, has a water content of 70-80%. In this state, paper sheet is just strong enough to be drawn off from the wire and transfered to the presses. The presses are heavy rolls to which pressure is applied, and through which the sheet passes, carried on the felt. There may be a number of presses. These presses may be straight where the top roll contacts the top side of the sheet, or reverse where the sheet is guided in the opposite direction, so that the top roll contacts the underside, or wire side of the sheet.

As much water as possible must be removed in this section, since the cost of water removal in the next stage, the dryers, is much higher than that of pressing. The common figures of moisture content after the presses are about 55%-65%.

Drying section

Drying is performed by passing the paper over a succession of hollow metal cylinders. The paper sheet is usually held tightly against the cylinders by "dry felts". These cylinders are steam-heated and provided with arrangements to remove the condensate. The water evaporates as the sheet passes through the drying section. Finally the sheet leaves the dryer with a moisture content of about 4%-8%.

Steam-heated cylinders are the most common type of dryers in use. In some cases, especially where coating has to be dried or space is at a premium, high-velocity hot air hoods are used, often in conjunction with cylinders. Other types of dryers in use include infrared heaters, dielectric and microwave dryers.

Calenders

Before reeling up, the sheet is passed between finely ground steel rolls under pressure, to provide a fine surface finishing. These rolls are in sets called calender stacks and more than one may be provided.

Drive

The paper machine is usually driven by electric motors. The sheet stretches as it passes through the machine, so that the drive must be sectionalized to provide for increases in speed at different points. The difference in speed between adjacent sections is called the draw. Draw must be provided at various points and the dryers must be sectionalized. The usual arrangement is to have one prime mover, the main drive, driving a line shaft, from which the different sections are driven through cone pulleys by belts. The belts are positioned to control the draw by remotely controlled motors or similar devices. In modern machines, the drive itself is sectionalized: each section has its own motor, the speeds of the different sections are controlled at the defined ratios by electronic devices.

Finishing

Finishing process includes super-calendering, slitting and reeling, cutting, sorting, etc.

1.2 Paper Machine Control Problems

This section briefly reviews the modeling and control problems, and the advanced control techniques which have been used in the paper-making process.

1.2.1 Headbox control

There exist some technically interesting and economically important control problems in the headbox. Figure 1.2 shows a simplified schematic of a headbox cross section. Figure 1.3 shows a signal flow diagram for two important variables: jet velocity S and liquid level L, which are affected by the stock-inflow Q_i, and the air inflow V_i. For regulation purposes, the objective of headbox control system is to maintain S and L constant (Xia and Rao, 1992; Xia, et al., 1993).

The jet velocity S is a calculated variable, being proportional to the square root of the total head that is measured. The most important variable is the "rush/drag," i.e., the ratio of the jet speed to the wire speed. This ratio has considerable influence on the properties of the sheet. Variations in rush/drag also show up as variations in basis weight and moisture content, which in turn

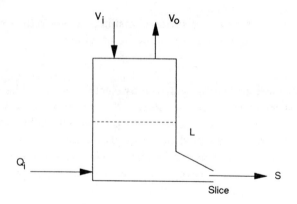

Figure 1.2: Headbox cross section

affect many other sheet properties. Rush/drag is also a calculated variable. The regulation of rush/drag and pond level are selected as controlled variables in advanced control systems due to the following reasons:

1) Rush/drag is a calculated variable and analog methods generally become inaccurate at high machine speeds;

2) Rush/drag and level interact each other, as shown in Figure 1.3. Unless decoupling control is used, the individual loop controllers will be difficult to to tune;

3) The process parameters, particularly those associated with blocks 1 and 2, depend strongly on the operating point, i.e., the machine speed. Blocks 3 and 4 are nonlinear. If a machine operates over various speeds, the traditional single-loop PID control is difficult to handle since it is generally tuned for one speed. No standard approach has been developed to solve these problems.

Headbox consistency control is important in that, together with jet velocity, it largely affects the dry basis weight. Control of consistency at the headbox can be a part of a feedforward scheme for basis weight control because of the dead time through the machine (40 seconds to 3 minutes). At the present time, no universal instrument is available for consistency measurement in the range of 0.5% to 1%, although a few research results on this issue have been reported (Brewster et al., 1970).

Headbox consistency can be controlled by an "open loop structure," i.e., without feedback, by calculating the material balance. This is based on an assumption about the retention of the Fourdrinier. Also needed for this material balance is a measure of the flow to the headbox. An alternative method

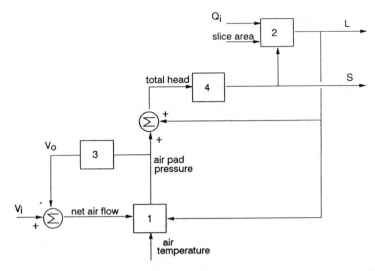

Figure 1.3: Headbox signal flow

of determining the flow rate is by multiplying jet velocity and slice open area. Absolute measurements of the latter contain some inaccuracies, particularly due to variations in cross-direction slice opening. Another factor required is the discharge coefficient.

According to Nader (1980) and Nader et al. (1979), the linear control system becomes unstable when the headbox is operated too far away from the controller design point.

1.2.2 Basis weight and moisture control

Basis weight and moisture content are the two primary controlled variables of paper machine. Most other paper properties depend on them. The problems associated with basis weight and moisture content control are interactions, dead time, and the nature of the disturbances.

The most common manipulated variable chosen to control basis weight is stock flow. This interacts strongly with moisture, however. An increase in stock flow will increase the consistency in headbox, and increase the drainage resistance of the web, and then increase the water flow rate with sheet. In the dryer section, an increase in water loading together with a thicker sheet will result in an increase in final moisture. This interaction is normally quite nonlinear.

The most common variable used to control the moisture is the dryer steam pressure. Although manipulation of this variable does not affect dry basis weight, sheet moisture content directly alters total basis weight. A decoupling system has, therefore, been proved to be valuable in basis weight and moisture content control.

The basis weight valve is situated in the front of the headbox, while the basis weight and moisture gauges are located near the reel. There is a variable process transport lag of 1 to 3 minutes between these two points, depending on machine speed. Thus, an advanced deadtime control algorithm may be required, such as predictive control or adaptive control with direct delay estimation. Experience has shown that this gives much better performance over traditional systems (Brewster, et al., 1970; Elnaggar, et al., 1992; Xia, et al., 1993).

Many variables affect the basis weight and moisture control, far too many to be considered individually in the control, including additives, machine speed, and dryer parameters (such as syphon operation, condensate thickness, ambient humidity, ventilation, felt condition, and felt tension). Few of these are measurable, and so, as with any control system, their effect must be lumped and compensated for in the design.

In basis weight and moisture control design, there are essentially four processes to be modeled:

1) stock flow/basis weight,
2) stock flow/moisture,
3) dryer steam pressure/moisture, and
4) dryer steam pressure/basis weight.

The fourth can be eliminated if dry basis weight is controlled rather than total basis weight. The elimination of this interaction, however, does not mean all coupling can be ignored.

A steady-state model relating dry basis weight to stock flow may be derived from a mass balance around the machine. The major disturbances to this system are errors in the consistency measurement which in turn is sensitive to fiber flow rate, stock temperature, pH, fiber proportions, etc.

The stock flow/moisture interaction is due to the effect of stock flow on water removal in the Fourdrinier, the press section, and the dryer. Model development is still difficult to reach the stage where interactions may be easily predicted, although some developments have taken place relating to the Fourdrinier (Schoeffler, et al., 1965; 1966), the presses (Wahlstrom, 1968), and dryers (Holm, 1969). The number of parameters involved in these models, however, make their present use uneconomic. Therefore empirical identification has been used to determine this interaction.

With the process models defined and identified as above, the decoupling controller algorithm may be found using standard matrix and transform analysis techniques. Because the dynamics are different for the responses to stock flow and to steam , unsteady-state decoupling is preferable with attention being paid to physical realizability.

Rough tuning of the algorithms is done using the identification results, as discussed above, for the existing process conditions. A table of such parameters is stored if a number of grades are run. Fine tuning may be accomplished for each loop by a single tuning parameter related to the closed-loop response specified in the controller design.

Box and Jenkins (1963) described an "adaptive quality control" technique which has been used in the design of a control strategy for basis weight. A closely allied approach was taken by Åström (1964) who used linear stochastic control theory and a minimum variance control criterion. The resulting optimal control was very sensitive to model parameters, and was subsequently modified. Rounsley (1968) used a modification of the model reference adaptive control approach to construct a predictive control algorithm, which acted based on the difference between the current process error and that predicted from past control actions. This was similar to Buhr's (1969) control structure, which utilized the results of Dahlin and Åström, but extrapolated ramp disturbances, filtered by choosing widely separated samples, to achieve prediction. Buhr also compared his algorithm with Rounsley's, Dahlin's, and Åström's considering response dynamics, stability, and sensitivity to model parameters as well as structure.

All of the above approaches are based on linear models, even though it is well recognized that nonlinearities do exist (e.g., moisture content time response to dryer steam pressure is different for an increase than for a decrease). The compensation methods for these nonlinearities have been developed, but few were reported to be implemented in real industry.

1.2.3 Cross Direction Control

For several reasons, it is economically quite important to produce a sheet which has a flat profile of properties in the cross machine direction (CD). Achievement of the desired degree of uniformity requires the correct combination of process design and control technology.

Dry basis weight is, of course, the most basic of the paper properties, and its CD profile is determined by the design and operation of the headbox. The key issue in solving CD control problem is how to estimate the "true" CD profile and the process response.

To obtain the dry basis weight, measurements of the total weight and the moisture content at each point across the sheet are required. This can be accomplished by means of on-line scanning gauges or by off-line profilers. With an on-line computer, the former can generally provide more useful information per reel than the off-line profiler even though the scan is contaminated with machine direction (MD) variations.

Because of the gauge response time and detector window size, there is a phase difference in the gauge output for the different scan directions. Since the "CD profile" is usually recorded only in one direction, conventional control systems find it difficult to deal with the phase shift. Therefore, an advanced control strategy is to be developed to solve this problem.

Automatic closed-loop control of CD basis weight has been proposed (Halouskova, 1991), the problem still appears due to two reasons (Brewster, et al., 1970). First of all, the digital filter lag is quite long (as much as 15 minutes). This long lag, together with relatively infrequent major disturbances, results in infrequent or slow control action. Secondly, the process response between slice screw adjustments and CD profile can involve up to a 60×60 matrix for wide machines. Although this matrix would be normally quite sparse and predominantly diagonal, there are a number of phenomena which add complications. Adjustment of one slice screw loads adjacent slice screws. Changing of the slice opening at one point tends to create cross flows. Slice screws almost invariably also have blacklash.

In addition, there is a limit to the adjustment that can be made on any screw. Thus, it is unable to efficiently overcome hydraulic distribution deficiencies in the headbox. Many alternative methods for the adjustment of the CD fiber flow have been conceived. However, the successful applications are few.

Because of the strong interaction among dry basis weight, moisture, and other variables, the control of basis weight will alleviate problems with these variables. Additional moisture problems usually arise due to CD nonuniformities in pressing and drying. Recently, there appears rapid development for CD adjustment equipment of these processes such as the controlled crown press roll and pocket ventilating systems.

1.2.4   Machine optimization

Even on the simplest paper machines, there exist at least four product quality variables. The full automatic economic optimization of paper machine operation has not been achieved yet. As we recognized, speed is probably the most important variable for economic optimization, and little has been reported on its use as a manipulated variable. Generally the problem is to select the product quality set points relative to the product specifications so that there is the correct balance between increased production and increased rejects (Brewster, 1969). This depends on good product quality regulation. Additionally, there are usually "soft" operational constraints which must not be violated. For example, if the moisture at the couch is too high, damage to the sheet structure ("crushing") can occur in the presses, resulting in an increased likelihood of sheet breaks or product defects.

There are normally more manipulated variables in paper-making than the controlled variables. For example, on a simple machine there could be the following variables (Table 1.1).

Table 1.1 Controlled and manipulated variables

Controlled variables	Manipulated variables
Basis weight	Stock flow
Moisture	Dryer steam
Strength	Refining energy
Production rate	Speed
	Press loading
	headbox consistency
	Rush/drag
	Stock blend
	etc.

In general, the relations among the manipulated variables, the disturbance variables, the controlled variables, as well as constraints, are nonlinear and interactive. There is consequently a major issue in modeling before optimization can be fully implemented. The gains could, however, be substantial.

Because of the process complexity and interactions, and the lack of on-line measurements for some important quality variables, a few difficulties remain in paper-making process operation. Design of advanced control systems has required investment in process analysis. This investment has produced the increasing knowledge of process (Brewster, et al., 1970).

1.2.5 Intelligent Control

If we classify the single variable control as level 1, the multivariable control as level 2 and economic optimizing control as level 3, the problems discussed above are essentially on a second level. The problems of production scheduling, planning, process monitoring and maintenance are on a still higher level, and we believe that more developments must occur at this level. These problems are "management control" problems, and the control system at this level is a "information management system." Fast, accurate, and near-optimal response to rapidly changing market and plant conditions is economically important and depends on systems of such type (Rao and Xia, 1994).

Paper machine operation is continuous, but has to shut down and start up periodly due to washing wire, changing parts (such as wire, felts, calenders, etc.). In some cases, a paper machine needs to produce different kinds or different grades of paper. Most paper mills, still heavily depends on the operators' experience during the shutdown, start-up, grade changing process operations as well as emergency handling, since the conventional control system cannot handle these ill-formulated problems.

The solution to these problems may require intelligent control systems. These systems must be able to deal with both symbolic reasoning and numerical computation, to communicate with any dedicated systems which perform level 2 functions, and to interface with process operators. Intelligent systems are developed by using artificial intelligence (AI) technology. Expert systems, artificial neural networks, fuzzy systems all belong to intelligent systems, and are the most commonly used in industrial process monitoring and control. There are five main reasons to use intelligent control system for paper machines (Rao, et al., 1991):

(a) The control problems in paper machine involve ill-structured and difficult to be formulated problems in which the mathematical modeling is not amenable, and purely algorithmic methods are difficult to use. However, AI techniques provide programming methodology for solving these ill-formulated engineering problems.

(b) In paper-making processes, operating conditions are frequently changed based on different product specifications. There exist many periodic operational procedures. Expert systems are suitable for use in such an environment.

(c) Paper-making process control always deals with uncertain and fuzzy information. Conventional control systems fail to handle such information. However, intelligent control systems can process imprecise information.

(d) The stochastic occurrence of operational faults requires emergency handling in paper-making processes. Past experience has shown that intelligent fault diagnosis systems are very powerful in dealing with such complex situations.

(e) Intelligent control is a new technological challenge, which may change the methodology of process control.

In summary, the paper machine plays a key role in paper-making process. The characteristics of paper machine control problems are: multivariable, time varying, distributed parameter, time delay, nonlinear, ill-formulated model, and stochastic process. Good control systems can improve the paper machine performance which will improve product quality, decrease production cost, increase productivity, improve operational safety, and protect the environment. In the following chapters, we will discuss process modeling and advanced control systems for paper machines.

1.3   References

Alsholm O., Schoeffler J.D. and Sullivan P.R. An on-line mathematical model of a fine paper machine. Presented at the ISA 20th Ann. Conf., 1965

Åström K.J. Control problems in paper making. Presented at the IBM Scientific Computing Symp., NY., 1964

Box G.E.P. and Jenkins G.M. Further contributions to adaptive quality control: simultaneous estimation of dynamics – non-zero costs. Dept. of Statistics, University of Wisconsin, Madison, Tech. report, 1963

Brewster D.B. Economic gains from improved quality control. Presented at TAPPI Annual Meeting, 1969

Brewster Donald B. and Bjerring Andrew K. Computer control in pulp and paper 1961-1969. Proc of the IEEE, 1970; 58:49-69

Buhr R.J.A. A practical optimal control design procedure applied to Rolland's number 8 paper machine. Presented at the ISA/CPPA Symp. on Pulp and Paper Process Control, Vancouver, Canada, 1969

Dalin E.B. Integrated control of fourdrinier machine. Presented at the 2nd International Symp. on Water Removal at the presses and Dryers, Mont Gabriel, Quebec, Canada, 1968

Elnaggar A., Dumont G.A and Elshafei A.L. Adaptive control with direct delay estimation. Proc of Control Systems' 92, Whistler, B.C. Canada, 1992, pp 13-17

Halouskova A., Karny M. and Nagy I. Adaptive cross-direction control of paper basis weight. Proc of the 11th triennial world congress of the IFAC, Tallinn, USSR, 1991, pp 199-204

Holm R.A. Dynamics of the drying process. Presented at the BPBMA 4th fundamental Res. Symp., Oxford, England, 1969

McGill R. Measurement and Control in Papermaking. Adam Hilger Ltd., 1980

Nader A.D. Dual model reference adaptive control on paper machine headboxes. Proc of 4th IFAC PRP Conf. Ghent, Belgium, 1980, pp 239-249

Nader A.D., Lebeau B., Gauthier J.P., Foulard C. and Ramaz A. New developments in multivariable control of paper machine headboxes. Presented at Int. Symp. on Paper Machine Headboxes, Montreal, Canada, 1979

Rao M. and Xia Q. Integrated distributed intelligent system for on-line monitoring and control of pulp processes. Canadian Journal of Artificial Intelligence 1994; Winter Issue:5-10

Rao M. and Ying Y. Intelligent engineering approach to pulp and paper process control. Proc of 77th CPPA Annual Meeting, Montreal, Canada, 1991, pp A195-A199

Rounsley R.R. Predictive control applied to a dryer section. Presented at 2nd International Symp. on Water Removal at the Presses and Dryers, Montreal, Canada, 1968

Schoeffler J.D. and Sullivan P.R. A model of sheet formation and drainage on a fourdrinier. Presented at the TAPPI Engrg. Conf., 1965

Xia Q. and Rao M. Fault-tolerant control of paper machine headboxes. J Proc. Contr. 1992; 2:171-178

Xia Q., Rao M., Shen X. and Zhu H. Adaptive control of a paperboard machine. Pulp and Paper Canada 1994; (in press)

Xia Q., Rao M., Shen X., Ying Y. and Zurcher J. Systematic modeling and decoupling control of a pressurized headbox. Proc. 1993 Canadian Conf. on Electrical and Computer Engineering, Vancouver, 1993, pp 962-965

CHAPTER 2
PROCESS DYNAMICS AND MODELING

Paper machines are very complex processes. Identification models, sometimes, cannot meet operational requirements. It is very difficult to develop paper machine dynamic models solely by theoretical mechanism analysis. In this chapter, a systematic modeling method for paper machines is proposed, which combines mechanism analysis with experimental method. The model structure and part of model parameters are determined from mechanism analysis. The remaining model parameters are obtained from experimental data. The modeling process for a pressurized headbox and a paper machine which produces condenser tissue (with an open headbox) is presented.

The presentation of this chapter is organized in such a way: first the modeling of a pressurized headbox is presented in Section 2.2; then the paper machine which produces the condenser tissue is divided into three major sections: headbox, wire and press, and drying. The process dynamics and modeling for these sections are discussed in Sections 2.3, 2.4 and 2.5, respectively. The overall paper machine model is obtained by integrating the models from all the three sections according to mass and information flows. The model test is presented in Section 2.6.

2.1   Introduction

The modeling techniques for paper machines can be mainly divided into two types (Xia and Sun, 1989): 1) Black box method. This the treats investigated paper machine as a black box. The process model is directly derived from input-output data. System identification belongs to this type. 2) Process mechanism analysis method. With this method, a paper machine is divided into several sections according to process mechanism. The model for each section is obtained independently based on mechanism analysis. The overall paper machine model is then obtained by integrating these models together according to material and signal flows.

By comparison, the black box method needs little understanding of the process mechanism. Thus, it is applicable to such processes where the process mechanism is very complicated or unknown. Its disadvantage is that the exciting signal must be applied continually. Thus normal production may be affected by the exciting signal. Moreover, it heavily relies on the on-line measurement of the process inputs and outputs that may not be always available for paper machines. The advantage of the mechanism analysis method is that process modeling can be conducted without affecting normal production. Using mechanism models, we can better understand some important process variables that are not measurable on-line, thus, we are able to optimize process operations. Other advantages of mechanism models are the generality and adaptability to the changes of plant configuration and operation sequence. However, a complete understanding of the process mechanism is required.

The paper-making process has the following characteristics:

1) It is a complicated heat and mass transfer process. The process mechanism is not completely known yet. The results provided by different investigators are often contradictory to each other.

2) It is a highly nonlinear and distributed parameter system.

There has been considerable research progress on paper machine modeling. So far, the models used for controller design are usually obtained by system identification (Gentile, 1974). There are a number of significant results on mechanism analysis, such as heat and mass transfer in the drying section (Baines, 1973; Lemaitre, et al., 1980), water removal in the wire and press section (Anderson and Back, 1981; Codieux, 1983). However, these researches provided only partial differential equations which describe the mass and heat transfer mechanisms. A few unknown parameters are often involved in the equations. The results can be applied only for qualitative analysis, and very few were used for analysing the whole process behaviour (Lemaitre, et al., 1980). To solve the modeling problem, Xia and Sun (1989) proposed a method to combine the mechanism analysis and on-site test in order to develop a dynamical model for paper machines. The method has been applied to controller design and given improved control performance.

Based on the discussion above, it is very difficult to develop a model for paper machine solely by theoretical mechanism analysis. A more practical and efficient technique is to combine mechanism analysis with experimental methods. Using mechanism analysis, we can determine the model structure and some model coefficients. The unknown model coefficients are determined by on-site experiments (Xia, 1989).

This chapter is mainly devoted to the modeling of a paper machine which produces super thin condenser tissue of basis weight $8\text{-}10g/m^2$. Since the on-line measurements of the basis weight and moisture content of such thin tissue are not available, model-based prediction is required in order to implement basis weight and moisture content control. A number of process variables are measured for the purpose of quality prediction. They are the opening of pulp flow aperture, the flow rate of thick stock, the consistencies of thick and diluted stocks, the vacuums of suction boxes, the steam pressure in the drying section, as well as the surface temperatures of web drying cylinders. The process model must give the relationship between product quality and these variables. A black box model using system identification cannot meet this requirement.

A feasible modeling technique is to use a simple model with adequate accuracy. The model structure is so designed that better quality prediction performance can be achieved. A new concept, namely *equivalent cylinder* in the drying section, is proposed for this purpose.

The investigated paper machine uses an opened headbox to distribute the diluted pulp evenly and steadily onto the wire. However, many modern high speed paper machines use a pressurized headbox. In order that the results presented in this chapter are applicable to most commonly used paper machines, the modeling technique for a pressurized headbox which is from an existing pulp mill is also included.

2.2 Pressurized Headbox

Pressurized headboxes are used in most modern high-speed paper machines. The headbox discussed here is a volume chamber of 232 inches wide, 57 inches deep and 48 inches high. Its main purpose is to distribute the water-fiber suspension onto the Fourdrinier wire as evenly and steadily as possible. Inside the headbox is a rectifier which provides even distribution of pulp fiber onto the Fourdrinier wire.

The pulp stock is pumped into the headbox by a fan pump, and flows onto the wire through a slice lip as a result of the effect of total head. The air is

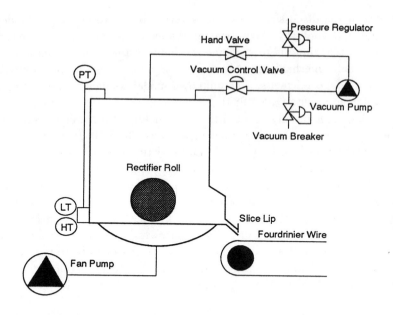

Figure 2.1: Flow chart of the pressurized headbox

circulating in the air chamber of headbox using a vacuum pump. The total stock flow rate into the headbox is controlled by fan pump speed. However, the change of fan pump speed influences only the white water flow but not thick stock flow. A ball valve (basis weight valve) is installed in the suction side of fan pump to adjust the fiber flow into the headbox. The liquid flow rate onto the Fourdrinier wire is affected by both total head and the opening of a manually adjustable slice lip. The slice lip opening should be adjusted so that a desired total head and constant stock consistency in the headbox can be held at a specified product basis weight. The air pressure in the headbox is controlled by a vacuum valve located in the suction side of the vacuum pump.

Figure 2.1 is a schematic diagram of the headbox. There are two controlled variables (outputs), i.e., the liquid level and the total head (or the pressure in the padding air chamber), and two manipulating variables (inputs), i.e., the fan pump speed and the vacuum air valve opening. The liquid level is controlled by manipulating fan pump speed. The total head is controlled by manipulating the vacuum air valve opening.

The pressurized headbox is inherently a second-order system. A combined modeling method of mechanism analysis and experimental tests is usually more effective. In this section, we first develop the model structure of the pressurized headbox using mass balance axiom and the ideal gas axiom, then the unknown parameters involved in the model are determined by step response tests on a real plant.

2.2.1   Modelling by theoretical analysis

The pressure in padding air chamber of the headbox is relatively low (-7∼15 inches water above atmospheric pressure). It is reasonable to assume that the air in the headbox is an ideal gas. Thus, we can apply the ideal gas axiom

$$m = M\frac{PV}{RT} \tag{2.1}$$

where m is the amount of air in the chamber, M is the equivalent molecular weight of air, P is pressure, V is the volume of the padding air chamber, T is the absolute temperature of the headbox, and R is the ideal gas constant.

By assuming the temperature in the headbox to be constant and defining a constant K_{c1} as

$$K_{c1} = \frac{M}{RT} \tag{2.2}$$

equation (2.1) can be rewritten as

$$m = K_{c1}PV \tag{2.3}$$

Taking derivative on both sides of equation (2.3) gives

$$\frac{dm}{dt} = K_{c1}P_0\frac{dV}{dt} + K_{c1}V_0\frac{dP}{dt} \tag{2.4}$$

where P_0 and V_0 are steady state values of pressure and chamber volume in the headbox, respectively.

Denoting liquid level as L, the change of vacuum air valve opening as u_2, the equivalent air flow resistance as R_2, the gain of air flow rate to vacuum air valve as K_{v2} and the cross-section area of the headbox as A, respectively, it is obvious that

$$\frac{dm}{dt} = -K_{v2}u_2 - \frac{\Delta P}{R_2} \tag{2.5}$$

$$\Delta V = -A\,\Delta L \tag{2.6}$$

where " Δ " denotes the deviation of variables from their steady state values.

Substituting equations (2.5) and (2.6) into (2.4) gives

$$-K_{v2}u_2 - \frac{P}{R_2} = -K_{c1}P_0A\frac{dL}{dt} + K_{c1}V_0\frac{dP}{dt} \tag{2.7}$$

Taking Laplace transformation on both sides of the equation, we obtain

$$(K_{c1}V_0 s + \frac{1}{R_2})P(s) = -K_{v2}U_2(s) + K_{c1}P_0 AL(s)s \tag{2.8}$$

The balance of liquid in the headbox gives

$$A\frac{dL}{dt} = Q_{in} - Q_{out} \tag{2.9}$$

where Q_{in} and Q_{out} are liquid flow rates into and out of headbox. Denoting the gain of liquid level to fan pump speed as K_{v1} and the equivalent flow resistance of the slice lip as R_1, obviously, we obtain

$$\triangle Q_{in} = K_{v1}u_1 \tag{2.10}$$

$$\triangle Q_{out} = \frac{\triangle P + \triangle L}{R_1} \tag{2.11}$$

where u_1 represents fan pump speed change. Substituting equations (2.10) and (2.11) into equation (2.9) gives

$$A\frac{dL}{dt} = K_{v1}u_1 - \frac{P + L}{R_1} \tag{2.12}$$

Taking Laplace transformation on both sides of equation (2.12) generates

$$AL(s)s = K_{v1}U_1(s) - \frac{1}{R_1}(P(s) + L(s)) \tag{2.13}$$

From equations (2.8) and (2.13), we can obtain the following transfer function model

$$
\begin{aligned}
L(s) &= \frac{K_{v1}K_{c1}R_1 R_2 V_0 s + K_{v1}R_1}{K_{c1}R_1 R_2 V_0 As^2 + (R_1 A + K_{c1}R_2 V_0 + K_{c1}R_2 P_0 A)s + 1}U_1(s) \\
&\quad + \frac{K_{v2}R_2}{K_{c1}R_1 R_2 V_0 As^2 + (R_1 A + K_{c1}R_2 V_0 + K_{c1}R_2 P_0 A)s + 1}U_2(s)
\end{aligned} \tag{2.14}
$$

$$
\begin{aligned}
P(s) &= \frac{K_{v1}K_{c1}R_1 R_2 P_0 As}{K_{c1}R_1 R_2 V_0 As^2 + (R_1 A + K_{c1}R_2 V_0 + K_{c1}R_2 P_0 A)s + 1}U_1(s) \\
&\quad - \frac{K_{v2}R_2(R_1 As + 1)}{K_{c1}R_1 R_2 V_0 As^2 + (R_2 A + R_1 K_{c1}V_0 + R_1 K_{c1}P_0 A)s + 1}U_2(s)
\end{aligned} \tag{2.15}
$$

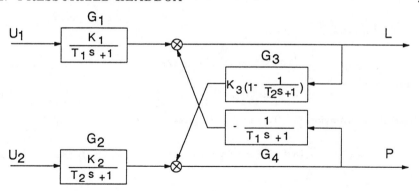

Figure 2.2: Block diagram of the pulp machine headbox

The total head H in the headbox is the summation of the liquid level and air pressure

$$H = L + P \tag{2.16}$$

From equations (2.14)-(2.16), we obtain

$$
H(s) = \frac{(K_{v1}K_{c1}R_1R_2V_0 + K_{v1}K_{c1}R_1R_2P_0A)s + K_{v1}R_1}{K_{c1}R_1R_2V_0As^2 + (R_1A + K_{c1}R_2V_0 + K_{c1}R_1P_0A)s + 1}U_1(s)
$$
$$
- \frac{K_{v2}R_1R_2As}{K_{c1}R_1R_2V_0As^2 + (R_1A + K_{c1}R_2V_0 + K_{c1}R_2P_0A)s + 1}U_2(s)
\tag{2.17}
$$

So far, we have developed a theoretical model for the pressurized headbox. There are four unknowns in the model: K_{v1}, K_{v2}, R_1, and R_2. These constants can be determined using data from bump tests. To describe the dynamics of the headbox more clearly, we will develop an alternative model by block diagram analysis method.

2.2.2 Modeling by block diagram analysis

Obviously, in the mechanism analysis method, the transfer function from liquid level to fan pump speed is of first-order if there were no influence of air pressure. Similarly, the transfer function from the pressure in air chamber to vacuum valve is also of first-order if there were no influence of liquid level. The liquid level does not have steady state effect on air pressure, and neither does the air pressure on total head. According to these observations, the block diagram of headbox can be depicted as shown in Figure 2.2.

From Figure 2.2 , we have

$$L(s) = G_1(s)U_1(s) + G_4(s)P(s) \tag{2.18}$$

$$P(s) = G_2(s)U_2(s) + G_3(s)L(s) \tag{2.19}$$

Substituting equation (2.19) into (2.18) gives

$$L(s) = G_1(s)U_1(s) + G_4G_2(s)U_2(s) + G_4(s)G_3(s)L(s) \tag{2.20}$$

or

$$
\begin{aligned}
L(s) &= \frac{G_1(s)}{1 - G_4(s)G_3(s)}U_1(s) + \frac{G_4(s)G_2(s)}{1 - G_4(s)G_3(s)}U_2(s) \\
&= \frac{K_1(T_2s + 1)}{T_1T_2s^2 + (T_1 + T_2 + K_3T_2)s + 1}U_1(s) \\
&\quad + \frac{K_2}{T_1T_2s^2 + (T_1 + T_2 + K_3T_2)s + 1}U_2(s)
\end{aligned}
\tag{2.21}
$$

Substituting equation (2.21) into (2.19) and rearranging it gives

$$
\begin{aligned}
P(s) &= \frac{G_3(s)G_1(s)}{1 - G_4(s)G_3(s)}U_1(s) + \frac{G_2(s)}{1 - G_4(s)G_3(s)}U_2(s) \\
&= \frac{K_1K_3T_2s}{T_1T_2s^2 + (T_1 + T_2 + K_3T_2)s + 1}U_1(s) \\
&\quad - \frac{K_2(T_1s + 1)}{T_1T_2s^2 + (T_1 + T_2 + K_3T_2)s + 1}U_2(s)
\end{aligned}
\tag{2.22}
$$

Comparing equations (2.21) and (2.22) with equations (2.15)-(2.17), it is found that the two methods give exactly the same transfer function models if the following equalities hold

$$K_1 = K_{v1}R_1, \qquad T_1 = AR_1 \tag{2.23}$$

$$K_2 = K_{v2}R_2, \quad T_2 = K_{c1}V_0R_2, \quad K_3 = P_0A/V_0 \tag{2.24}$$

Equations (2.21) and (2.22) show that due to the interactions, the open loop poles of the headbox are no longer $\frac{1}{T_1}$ and $\frac{1}{T_2}$, but the roots of the polynomial equation $T_1T_2s^2 + (T_1 + T_2 + K_3T_2)s + 1 = 0$. The effect of K_3 on the open-loop poles will be discussed in detail later.

2.2.3 Model coefficient determination

A series of bump tests have been performed on the headbox. During the tests, both the fan pump and the vacuum valve are placed in manual. Step changes are introduced into each of the final elements (fan pump and vacuum valve) while the other is held constant. The recorded data of the air pressure and liquid level are used to determine the unknown parameters in the transfer function model developed before.

It has been shown that the pressurized headbox is an inherently non-oscillatory second-order system. There is a simple procedure to determine the transfer function of such systems

$$W(s) = \frac{K}{W_d(s)} = \frac{K}{T^2 s^2 + 2\xi T s + 1} \tag{2.25}$$

This procedure is given as follows:

Step 1 Obtain the required data which include the time response of output $y(t)$, the table $y_4^* \sim \lambda$ and the input step change magnitude x_0;

Step 2 Compute $K = y(\infty)/x_0$;

Step 3 Compute standardized output $y^*(t) = y(t)/y(\infty)$;

Step 4 Compute time t_7 corresponding to $y^*(t_7) = 0.7$ using interpolating method;

Step 5 Compute $t_4 = t_7/3$ and determine $y_4^* = y^*(t_4)$ using interpolating method;

Step 6 Compute λ according to the table $y_4^* \sim \lambda$ using interpolating method;

Step 7 Compute T and ξ using the following equations

$$T = \frac{t_7}{2.4}\sqrt{1 - \lambda} \qquad \xi = \frac{1}{\sqrt{1 - \lambda}} \tag{2.26}$$

From equations (2.14), (2.15) and (2.17), we can see that only the transfer function of the liquid level to vacuum valve has the same form as equation (2.25). Therefore, the data of liquid level from the bump tests are used to calculate the unknown parameters in the headbox dynamic model.

From the step response of liquid level when vacuum valve position changes, the real time data are obtained as: $y(0) = 66.8$, $y(\infty) = 81.5$, $x(0) = 10$ ($52.6 \rightarrow 62.6$). The standardized output is

$$y^*(t) = \frac{y(t) - y(0)}{y(\infty) - y(0)} \tag{2.27}$$

It is calculated using the above procedure that $y(t_7) = 79.1$, $t_7 = 3.18$ (minutes), $t_4 = t_7/3 = 1.06$ (minutes), $y(t_4) = 71.64$, $y^*(t_4) = 0.3167$, $\lambda =$

0.9726. Using equation (2.26), we obtain $T = 1.281646$, $\xi = 5.170877$. The denominator of the transfer function is

$$W_d(s) = 0.03945655s^2 + 2.645s + 1 \tag{2.28}$$

or

$$W_d(s) = (0.015s + 1)(2.63s + 1) \tag{2.29}$$

The unknown constant $K_{v1}R_1$ is the steady state gain of liquid level to fan pump speed, and K_2R_1 is that of both air pressure and liquid level to vacuum valve position. They can be directly computed from the steady state data

$$K_{v1}R_1 = \frac{1}{2}\left(\frac{86 - 74.5 - 8}{2} + \frac{74.5 - 67.5}{4}\right) = 1.75 \tag{2.30}$$

$$\begin{aligned} K_{v2}R_2 &= \frac{1}{4}\left(\frac{40 - 24.2}{62.6 - 52.6} + \frac{80.8 - 67}{62.6 - 52.6} + \frac{41.6 - 24}{62.6 - 52.6} + \frac{81.2 - 66.8}{62.6 - 52.6}\right) \\ &= 1.54 \end{aligned} \tag{2.31}$$

$K_{c1}R_2$ and R_1 are computed by comparing the coefficients of the denominator in equation (2.14) with the coefficients in equation (2.28)

$$K_{c1}R_1R_2 = 0.09345 \tag{2.32}$$

$$R_1A + K_{c1}R_2V_0 + K_{c1}R_2P_0A = 2.645 \tag{2.33}$$

Solving R_1 and $K_{c1}R_2$ from equations (2.32) and (2.33) and substituting $K_{v1}R_1$, $K_{v2}R_2$, R_1 and $K_{c1}R_2$ into equations (2.14), (2.15) and (2.17), we finally obtain the transfer function model of the pressurized headbox

$$Y_1(s) = A_1(s)U(s) \tag{2.34}$$

$$A_1(s) = \begin{pmatrix} \frac{6.504s}{(2.63s+1)(0.015s+1)} & \frac{-1.54(0.1024s+1)}{(2.63s+1)(0.015s+1)} \\ \frac{1.75(0.1788s+1)}{(2.63s+1)(0.015s+1)} & \frac{1.54}{(2.63s+1)(0.015s+1)} \end{pmatrix} \tag{2.35}$$

where $Y_1(s) = (P(s)\ L(s))^T$, $U(s) = (U_1(s)\ U_2(s))^T$, or

$$Y_2(s) = A_2(s)U(s) \tag{2.36}$$

$$A_2(s) = \begin{pmatrix} \frac{1.75(3.8953s+1)}{(2.63s+1)(0.015s+1)} & \frac{-0.1578s}{(2.63s+1)(0.015s+1)} \\ \frac{1.75(0.1788s+1)}{(2.63s+1)(0.015s+1)} & \frac{1.54}{(2.63s+1)(0.015s+1)} \end{pmatrix} \tag{2.37}$$

where $Y_2(s) = (H(s)\ L(s))^T$.

The comparison of the model output and the real process output is shown in Figures 2.3 and 2.4. The model output fits the real process output data quite well.

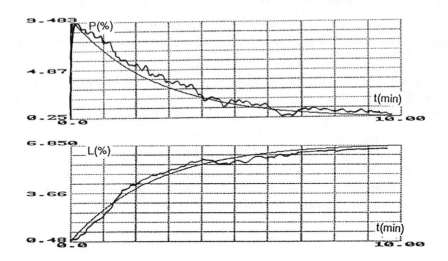

Figure 2.3: Simulation when fan pump speed changes by 4

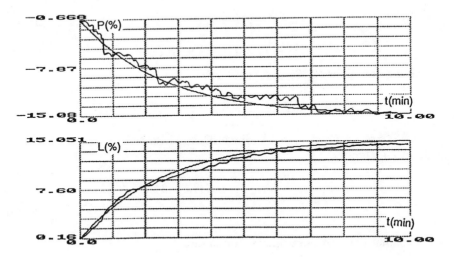

Figure 2.4: Simulation when vacuum valve position changes by 10

2.2.4   System characteristics analysis

A number of important observations concerning the characteristics of the pressurized headbox can be made directly from the model obtained above:

(1) From equations (2.21) and (2.22), we find that the open-loop poles of the headbox are altered by the interactions between the two output variables P and L. The physical meaning of K_3 is the amount of air pressure change caused by unit change of liquid level. K_3 can be calculated from the ideal gas axiom

$$P_1 V_1 = P_2 V_2 \tag{2.38}$$

The liquid level change results in air chamber volume change. In the normal operating condition, the air pressure is $P_0 = 406.794$, and the height of air chamber is $N = 31$ inches. That is

$$K_3 = P_0 \frac{1}{N} = 13.12 \tag{2.39}$$

Since K_3 is much larger than 1, it can be expected that the open-loop poles must be quite different from that of the original separated loops. The quantitative analysis is given as follows.

Denoting the open-loop poles of the headbox by $-\frac{1}{T_1'}$ and $-\frac{1}{T_2'}$, from equations (2.21) and (2.22) we have

$$T_1' T_2' = T_1 T_2 \tag{2.40}$$

$$T_1' + T_2' = T_1 + T_2 + K_3 T_2 \tag{2.41}$$

Solving equations (2.40) and (2.41) gives

$$T_1' - T_2' = \sqrt{T_1^2 + (1 + K_3)^2 T_2^2 + 2(K_3 - 1)T_1 T_2} \tag{2.42}$$

where we have assumed $T_1' > T_2'$. Since $K_3 = 13.12 \gg 1$, equation (2.42) can be approximated by

$$T_1' - T_2' \approx \sqrt{T_1^2 + K_3^2 T_2^2 + 2K_3 T_1 T_2}$$
$$= T_1 + K_3 T_2 \tag{2.43}$$

Solving equations (2.41) and (2.43), we obtain

$$T_1' = T_1 + (K_3 + 0.5)T_2 \tag{2.44}$$

$$T_2' = 0.5 T_2 \tag{2.45}$$

It is obvious from equations (2.44) and (2.45) that $T_1 \gg T_2$.

In an alternative way, solving equations (2.40) and (2.41) gives

$$T_1 - (1 + K_3)T_2 = \pm\sqrt{T_1'^2 + T_2'^2 - 2(2K_3 + 1)T_1'T_2'} \qquad (2.46)$$

Because T_1 and T_2 are positive real number, the following condition must be satisfied

$$T_1'^2 + T_2'^2 - 2(2K_3 + 1)T_1'T_2' \geq 0 \qquad (2.47)$$

or

$$T_1' > (2K_3 + 1)T_2' = 27T_2' \qquad (2.48)$$

That is to say one of the poles is at least 27 times as large as another one.

(2) Although the headbox is inherently a second-order system, it can be reduced to a first-order model because of the great difference between these two poles. The simplified first-order model is given as

$$Y_1(s) = F_1(s)U(s) \qquad (2.49)$$

$$Y_2(s) = F_2(s)U(s) \qquad (2.50)$$

where $Y_1(s)$, $Y_2(s)$ and $U(s)$ are the same as in equations (2.34) and (2.36). $F_1(s)$ and $F_2(s)$ are given as

$$F_1(s) = \begin{pmatrix} \frac{6.504s}{(2.63s+1)} & \frac{1.54(0.1024s+1)}{(2.63s+1)} \\ \frac{1.75(0.1788s+1)}{(2.63s+1)} & \frac{1.54}{(2.63s+1)} \end{pmatrix} \qquad (2.51)$$

$$F_2(s) = \begin{pmatrix} \frac{1.75(3.895s+1)}{(2.63s+1)} & \frac{-0.1578s}{(2.63s+1)} \\ \frac{1.75(0.1788s+1)}{(2.63s+1)} & \frac{1.54}{(2.63s+1)} \end{pmatrix} \qquad (2.52)$$

(3) The operators observed a "very strange" phenomenon in operation: the step change of fan pump speed will cause the pressure in air chamber to change abruptly. However, from equations (2.22) and (2.51) the reason of this phenomenon is very clear: since the system is very close to first-order and K_3 is quite large, the response of air pressure to the change of fan pump speed has an almost abrupt change with the gain $K_1K_3T_2/T_1'$.

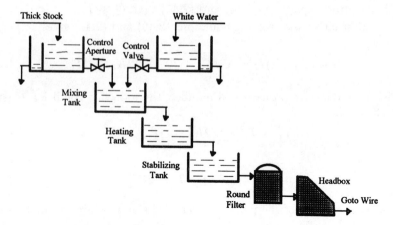

Figure 2.5: Layout of the open headbox section

2.3 Open Headbox

The open headbox section in the investigated paper machine consists of a high-level tank, mixing tank, heating tank, stabilizing tank and headbox (as depicted in Figure 2.5). The thick stock flows out of the high-level tank with its flow rate controlled by an aperture, and then diluted in the mixing tank by white water. The diluted stock flows onto the fourdrinier wire through the slice of the headbox. Each container can be considered as a first order process. Therefore, the headbox section is a serial connection of four first-order systems with time delays. However, among the four first-order systems, the time constant of the headbox is significantly larger than that of the other systems. Therefore, the headbox section is usually approximated by a first-order system with time delay for the convenience of controller design.

2.3.1 Static model

From the total mass and fiber balance, we have

$$G_{20} = G_{w0} + G_{10} \tag{2.53}$$

$$G_{20}C_{20} = G_{w0}C_{w0} + G_{10}C_{10} \tag{2.54}$$

$$G_{20} = G'_{w0} + Bw_0 Sp_0 L(1 - Ms_0)/(1 - Ms_{10}) \tag{2.55}$$

$$G_{20}C_{20} = G'_{w0}C'_{w0} + Bw_0 Sp_0 L(1 - Ms_0) \tag{2.56}$$

where the subscript "0" represents steady states of process variables and

G_1:   thick stock flow rate (t/min);
G_2:   diluted stock flow rate onto the wire (t/min);
G_w: circulating white water flow rate (t/min);
G'_w: white water flow rate drained from the wire (t/min);
C_1:   thick stock consistency (%);
C_2:   diluted stock consistency (%);
C_w: circulating white water consistency (%);
C'_w: drained white water consistency (%);
Bw: basis weight on reel (g/m²);
Ms: moisture content on reel (%);
Ms_1: moisture content after the second press (%);
Sp: machine speed (m/min);
L: width of paper web across machine (m).

The steady states of process variables C_{20}, C_{10}, C'_{w0}, Bw_0, Ms_0, Sp_0, L and C_{w0} can be obtained from the on-site tests. G_{20}, G_{10} and G_{w0} are obtained from solving equations (2.53)-(2.56).

Experiments have shown that the consistency of the white water drained from the wire is proportional to the pulp consistency in the flowbox, that is

$$C'_w = k_1 C_2 \tag{2.57}$$

The difference between the consistency of circulating white water and the consistency of drained white water is effected by wire wash water. It is reasonable to assume that there is a proportional relation between these two consistencies

$$C_w = k_2 C'_w \tag{2.58}$$

Thus

$$C_w = k_1 k_2 C_2 = k C_2 \tag{2.59}$$

where the proportional coefficient $k = 0.2$ is obtained from normal operating conditions.

Taking deviation on both sides of equation (2.53), (2.54) and (2.59), we obtain

$$\Delta G_2 = \Delta G_w + \Delta G_1 \tag{2.60}$$

$$G_{20}\Delta C_2 + C_{20}\Delta G_2 = C_{w0}\Delta G_w + G_{w0}\Delta C_w + G_{10}\Delta C_{10} + C_{10}\Delta G_1 \tag{2.61}$$

$$\Delta C_w = k\Delta C_2 \tag{2.62}$$

The steady state gains of the model can be obtained from the above equations

$$\frac{\Delta C_2}{\Delta C_1} = \frac{G_{10}}{G_{20} - kG_{w0}} = 0.3818 \tag{2.63}$$

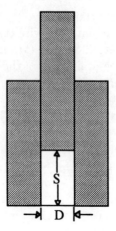

Figure 2.6: The thick stock control aperture

$$\frac{\Delta C_2}{\Delta G_1} = \frac{C_{10} - C_{20}}{G_{20} - kG_{w0}} = 1.696 \qquad (2.64)$$

$$\frac{\Delta C_2}{\Delta G_w} = \frac{C_{w0} - C_{20}}{G_{20} - kG_{w0}} = -1.226 \qquad (2.65)$$

The flow rate of thick stock is controlled by an aperture installed in the high-level tank, as depicted in Figure 2.6. Denoting the height of the pulp in the tank by H, the width of the aperture by D, and the opening height by S, the basic principle of fluid mechanics gives the following equation

$$dG_1 = k_g[2g(H - S)]^{\frac{1}{2}} D ds \qquad (2.66)$$

where k_g is a constant which reflects fluid mechanism of pulp flows out of the aperture.

Integrating the above equation gives

$$
\begin{aligned}
G_1 &= \int_0^s dG_1 \\
&= \frac{2}{3} k_g [2g(H - S)]^{\frac{1}{2}} D[H^{\frac{3}{2}} - (H - S)^{\frac{3}{2}}]/(H - S)^{\frac{1}{2}} \qquad (2.67)
\end{aligned}
$$

Substituting the steady state aperture opening and the thick stock flow rate into equation (2.65), we obtain k_g

$$k_g[2g(H - S)]^{\frac{1}{2}} D = 3.487 \times 10^{-3} \qquad (2.68)$$

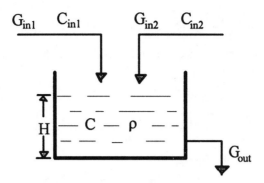

Figure 2.7: Principle diagram of a single tank

or

$$\Delta G_1 = 3.487 \times 10^{-3} \Delta S \qquad (2.69)$$

The steady state gains obtained so far are based on physical units of the process variables. To be processed by process control computer, these gains must be based on digital code. According to the measurement range of the sensors, the converted gains are obtained as follows:

$$(\frac{\Delta C_2}{\Delta C_1})_d = 0.8754, \qquad (\frac{\Delta C_2}{\Delta G_1})_d = 1.188, \qquad (\frac{\Delta C_2}{\Delta G_w})_d = -0.5586$$

$$(\frac{\Delta G_1}{\Delta S})_d = -0.2583, \qquad (\frac{\Delta G_w}{\Delta I_2})_d = 0.8754 \qquad (2.70)$$

where the subscript "d" represents digital code, and I_2 represents the control signal to white water control valve.

2.3.2 Dynamical model

The steady state gains of the headbox model have been obtained in Section 2.3.2. In order to determine the time constant, we first consider a single tank with two inflows and one outflow of the same density, as depicted in Figure 2.7.

In Figure 2.7, G_{in1} and G_{in2} are the flow rates of two inflows; C_{in1} and C_{in2} are the consistencies of two inflows; G_{out} is the flow rate of outflow; C and ρ are the consistency and density of the fluid in the tank. It is assumed that the fluid in the tank has uniform properties.

According to mass and fiber balance, we obtain the following relations

$$\frac{d(H\rho A)}{dt} = G_{in1} + G_{in2} - G_{out} \qquad (2.71)$$

$$\frac{d(H\rho AC)}{dt} = G_{in1}C_{in1} + G_{in2}C_{in2} - G_{out}C \tag{2.72}$$

where A is the cross section area of the tank. From the above equation, we have

$$\frac{dH}{dt} = \frac{G_{in1} + G_{in2} - G_{out}}{A\rho} \tag{2.73}$$

$$\frac{dC}{dt} = \frac{G_{in1}(C_{in1} - C) + G_{in2}(C_{in2} - C)}{A\rho H} \tag{2.74}$$

The mass conservation in steady state gives

$$G_{in10} + G_{in20} = G_{out0} \tag{2.75}$$

$$G_{in10}C_{in10} + G_{in20}C_{in20} = C_{out0}C_0 \tag{2.76}$$

The outflow rate and the liquid level in the tank are related by

$$\Delta G_{out} = \Delta H/R \tag{2.77}$$

where R is flow resistance.

Substituting equations (2.75) and (2.76) into (2.73) and (2.74) and taking the Laplace transformation of the resulting equations gives

$$G_{out}(s) = \frac{1}{T_g s + 1}(G_{in1}(s) + G_{in2}(s)) \tag{2.78}$$

$$
\begin{aligned}
C(s) = {} & \frac{1}{T_c s + 1}(k_{c1}G_{in1}(s) + k_{c2}G_{in2}(s) \\
& + k_{c3}C_{in1}(s) + k_{c4}C_{in2}(s) + k_{c5}G_{out}(s))
\end{aligned} \tag{2.79}
$$

with

$$T_g = A\rho R, \qquad T_c = A\rho H_0/(G_{in10} + G_{in20})$$

where k_{ci} (i=1,2,3,4,5) can also be obtained easily.

As mentioned above, the dynamics of the headbox section is a serial connection of a number of the first-order systems with the form of transfer function as shown in equations (2.78) and (2.79). In order to obtain a simplified equivalent first-order model for the headbox section, we apply step response method rather than complex model reduction techniques: i.e. simulating the serial connection of the first-order systems in computers; introducing step changes in inputs and observing the corresponding output response. The equivalent time constants of the model thus can be obtained

$$T_{g1} = 2.5 min, \quad T_{gw} = 2.5 min, \quad T_{c2} = 3.7 min \tag{2.80}$$

The transportation lag of the pipe lines in the headbox section is

$$\tau = 27 sec \tag{2.81}$$

Selecting sampling period $T_s = 20 sec$, the discrete time model of the headbox section can be obtained by using the steady state gains given in equation (2.70), time constants in (2.80) and time delay in (2.81)

$$G_1(k+1) = 0.8667 G_1(k) - 0.03443 S(k-1) \tag{2.82}$$

$$G_w(k+1) = 0.8667 G_w(k) - 0.6877 I_2(k-1) \tag{2.83}$$

$$C_2(k+1) = 0.9099 C_2(k) + 0.1069 G_1(k) - 0.05033 G_w(k) + 0.07881 C_1(k-2) \tag{2.84}$$

It should be pointed out that the system dynamics has been greatly simplified in order to obtain the above first-order model. Model errors are inevitable. However, since the steady state gains are accurately calculated from mass balance, the model will be accurate enough for controller design.

2.4 Wire and Press

For every ton of paper made on a paper machine, about 100 tons to 300 tons of water are used by the machine at the slice. The basic functions of the wire and press section are to remove up to 97% of the water and to form paper sheet. The water removal devices include: 34 table rolls, 12 suction boxes and 2 stages of press. Water removal in this section is a very complex mass transfer process. It is affected by quite a few factors, such as Kappa number (or freeness), consistency and temperature of the pulp on the wire and the worn state of the wire. Though there have been a number of reports on the mechanism of water removal, only empirical equations were given and these were valid only for some specific processes. On-site tests and experiments are required to develop a simple and accurate model.

In order to simplify the modeling process and obtain a suitable model for controller design, the following assumptions are made:

(1) The free drainage water removal on the wire between two table rolls is equal to water removal at the rolls. In this way, the wire can be discretized in the interval of a single roll.

(2) The water removed in the presses does not contain any fiber.

2.4.1   Water removal at table roll

Kobah suggested that water removal at table rolls can be represented by
(Kugushev, 1982)

$$W = \frac{4rK_1gh}{S_p} \tag{2.85}$$

where r is the radius of roll; g gravity; h the thickness of pulp web on wire;
S_p wire speed; K_1 a constant which depends on temperature, consistency and
ratio resistance of pulp, with

$$K_1 \propto 1 - 0.09(SR - 96.5) \tag{2.86}$$

and

$$K_1 \propto (\frac{\bar{C} - C}{C})^{n_1} \tag{2.87}$$

where $\bar{C}$ is the consistency of pulp web formed on wire and is considered to be
the consistency of paper sheet when it leaves the wire; SR is Kappa number;
n_1 is a constant.

In order to improve the accuracy of the water removal mechanism given by
equations (2.85)-(2.87), we combine them into a single expression

$$W_i = K_i(\frac{\bar{C} - \bar{C}_{ri}}{\bar{C}_{ri}})^{n_1}(\bar{h}_i)^{n_2}(S_p)^{-n_3}(1 - 0.09(SR - 96.5))$$

$$i = 1, 2, \cdots, 34 \tag{2.88}$$

where W_i is the amount of water removed from unit length of the ith roll per
unit time; $\bar{C}_{ri} = \frac{1}{2}(C_{ri} + C_{r(i-1)})$ and $\bar{h}_i = \frac{1}{2}(h_i - h_{i-1})$ are the arithmetic
average of the pulp consistency and web thickness on the wire between the
$(i-1)th$ and ith table rolls, which can be calculated according to mass balance.

The initial pulp web thickness h_0 can be calculated from the pulp flow rate
on the wire, G_2 thickness is given as

$$G_2 = h_0 S_p L \tag{2.89}$$

where S_p is machine speed; L is width of pulp web across the machine. The
thickness at the outlet of the ith roll is

$$h_i = h_{i-1} - W_i/S_p, \qquad i = 1, 2, \cdots, 34 \tag{2.90}$$

and the corresponding consistency C_{ri} is obtained from dry fiber balance

$$G_2C_2/L = \sum_{k=1}^{i} C_{wk}W_k + C_{ri}(G_2/L - \sum_{k=1}^{i} W_k) \tag{2.91}$$

where C_2 is the pulp consistency in headbox; C_{wi} is the consistency of white water drained at the ith roll.

Table 2.1 shows the amount and consistency of white water drained from every two rolls.

Table 2.1: Water removed at table rolls

Table rolls No.	Amount of white water (ml/min)	Consistency of white water (g/l)
1~ 2	8426	0.2736
3~ 4	9170	0.4446
5~ 6	8240	0.6266
7~ 8	5380	0.6556
9~ 10	4337	0.6426
11~ 12	3933	0.5310
13~ 14	3445	0.5620
15~ 16	3105	0.6196
17~ 18	2307	0.6890
19~ 20	2542	0.6150
21~ 22	2366	0.6180
23~ 24	2168	0.6186
25~ 26	2032	0.5478
27~ 28	1953	0.6894
29~ 30	1859	0.5742
31~ 32	1784	0.5652
33~ 34	1734	0.5534

Using the data given in the table, model regression using equation (2.88) generates

$$n_1 = 0.36, \qquad n_2 = 0.74 \tag{2.92}$$

and operational experience suggested that

$$n_3 = 0.87 \tag{2.93}$$

2.4.2 Water removal in suction boxes

The water removal rate in unit length of a suction box was given by (Kugushev, 1982)

$$W_v = S_p h_v (1 - e^{-kB\sqrt{H_v}/(qS_p)}) \tag{2.94}$$

where H_v is vacuum of the suction box; h_v is the thickness of pulp web on surface of the suction box; q represents the amount of dry fiber in unit area; B is the width of suction box in machine direction; k is an unknown coefficient which depends on the filtering property of the surface of the suction box and is to be determined from the steady state process parameters. The vacuum distribution of the suction boxes is given in Table 2.2. Since the control system for the paper machine measures only the vacuum of one suction box, we could assume that the vacuums of all suction boxes change proportionally.

Table 2.2: Vacuum distribution in suction boxes

Suction box No.	Vacuum pressure (mmH_2O)	Suction box No.	Vacuum pressure (mmH_2O)
1	0	2	-5
3	35	4	60
5	67	6	44
7	-10	8	88
9	295	10	312
11	320	12	410

2.4.3 Water removal at presses

Since the on-line measurement of pin pressure of the presses is not available, we use a very simple method to deal with water removal at the presses. Experiments have shown that the variation of moisture contents into and out of the press section fits the following equation

$$\Delta M s_{out} = k_p \Delta M s_{in} \tag{2.95}$$

where $k_p = 0.38$; $\Delta M s_{in}$ and $\Delta M s_{out}$ represent the deviation of moisture content into and out of press section from their steady state values.

2.4.4 Lumped model

Based on the discussion above, the lumped model of the wire and press section can be obtained using the following procedure:

1) Using equations (2.88)-(2.91) to calculate the consistency and thickness of pulp web at every table roll;

2) Using equation (2.94), that is

$$W_{vi} = S_p h_{vi} \left(1 - e^{-kB_i \sqrt{H_{vi}}/(q_i S_p)}\right), \quad i = 1, 2, \cdots, 12 \tag{2.96}$$

and applying the method in a similar way to that used for table rolls to calculate the amount of removed water, the thickness of pulp web and the amount of dry fiber per unit area at every suction box. Since an unknown coefficient, k, is included in equation (2.94), we assume that all suction boxes have the same k value. This coefficient is determined by using the steady state value of moisture content of paper web into and out of the suction box section.

3) Introducing step changes to the flow rate of pulp onto the wire, the pulp consistency in headbox, the machine rate, the vacuum of suction boxes and Kappa number, respectively, and applying the above equations to calculate the corresponding basis weight and moisture content after the second press; Calculating the steady state gains from the date and transforming them into dimensionless gains that can be processed by computer.

4) Assuming the wire and press section is a pure time delay system. The time delay is

$$\tau = 18sec \qquad (2.97)$$

The obtained mathematical model is

$$\begin{aligned} Bw_1(k+1) &= 4.09G_1(k) + 1.491G_w(k) + 1.753C_2(k) - 8.543S_p(k) \\ &\quad +4.03 \times 10^{-4}SR(k) - 3.01 \times 10^{-3}H_v(k) \qquad (2.98) \end{aligned}$$

$$\begin{aligned} Ms_1(k+1) &= 5.847G_1(k) - 1.671C_2(k) + 2.842G_w(k) - 1.811S_p(k) \\ &\quad +1.181 \times 10^{-4}SR(k) - 7.272 \times 10^{-3}H_v(k) \qquad (2.99) \end{aligned}$$

2.5 Drying Section

The drying section is a very important unit of the paper making processes. Most of the mechanical properties of paper products are acquired from this stage. Improper operation of drying section may produce serious defects, such as curl, overdrying, blistering, dryer wrinkles, and so on.

The drying section in our industrial paper machine consists of 7 sheet dryer cylinders and 7 felt dryer cylinders. These dryers are mounted in two horizontal rows. Paper sheet passes through sheet dryers in a serpentine fashion. It is wrapped around one dryer in top row and then around another in the bottom row. The sheet is pressed tightly against the dryers by a heavy dryer felt. The dryers are rotated in synchronization to facilitate the passage of the sheet.

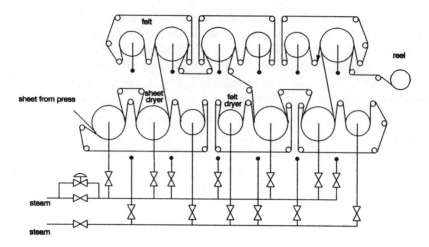

Figure 2.8: Flow diagram of the drying section

Heat is supplied by steam condensing inside the cylinder. The paper sheet travels back and forth between the two rows of dryers until it is dry. The sheet moisture content at the entrance of the drying section is about 60%-70%. At the end, the moisture content must be about 6%-8%. A diagram of the drying section is shown in Figure 2.8.

The drying section of the paper machine must provide each water molecule inside the wet sheet with enough energy to change from the liquid state to the vapor state, to break the chemical and/or mechanical bonds, and to overcome frictional resistance to liquid and vapor flow. This energy is in the form of heat, which raises the kinetic energy (temperature) of the water molecules, thus allowing them to break away from the sheet. The amount of energy is adjusted by a steam control valve which regulates the steam flow rate to the dryer cylinders.

In the following, a systematical modeling method for the drying section is introduced. Fick's second diffusion law and the mass/energy balance principle are applied. The steam supply system is simulated via an analogous electrical network to simplify the modeling processes. The concept of "equivalent dryer" is introduced to represent the dynamics of the drying section. The influence of all dryers on moisture is measured by the "equivalent cylinder" temperature. The introduction of the "equivalent dryer" makes the drying section model simpler, more accurate and robust.

2.5.1 Assumptions

Water removal in the drying section involves two basic physical processes, heat transfer and mass transfer. Heat is transferred from the steam to the wet sheet in order to provide the energy required to drive off the moisture. The moisture evaporates from the sheet and is then transferred from the sheet to the ambient medium by the mass transfer process. A number of hypotheses are made in order to obtain a workable dynamic model:

(1) The physical characteristics of sheet, such as thickness, temperature, moisture content and basis weight, are uniform across the machine. Only variations in the machine direction are considered.

(2) There is no relative motion between sheet and felt.

(3) All the internal phenomena inside sheet are described by the vapour pressure at sheet surface. The dynamic phenomena are neglected.

(4) Since the heat capacity of sheet is small, the temperature of sheet on dryer is assumed to be the temperature of the dryer cylinder.

2.5.2 Moisture content model

The water removal from wet sheet to ambient medium can be divided into two phases: moisture movement within wet sheet and from the sheet surface to the ambient medium. The moisture and vapour pressure gradients are the main causes of moisture diffusion.

The vapour pressure at the sheet surface depends on pulp characteristics, moisture content, temperature, etc. It can be described by

$$P_w = \begin{cases} P_s & Ms \geq M_{sc} \\ P_s e^{-k_f(1/Ms - 1/M_{sc})} & otherwise \end{cases} \tag{2.100}$$

where P_w is vapour pressure on sheet surface, P_s is the saturated vapour pressure at sheet temperature, M_{sc} is the critical moisture content of the wet sheet, k_f is a constant which depends on pulp characteristics. Here, we choose $M_{sc} = 0.51$ and $k_f = 0.048$.

According to the mechanism of water removal, we investigate water removal for two typical positions: sheet on the dryer cylinder and sheet in the draw (Figure 2.9).

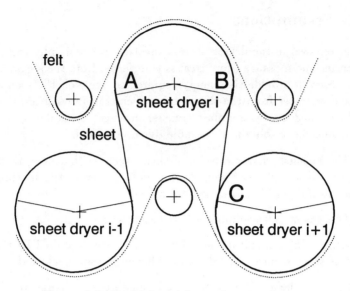

Figure 2.9: Typical positions of paper sheet

Sheet on the dryer cylinder ($\widehat{AB}$)

When a sheet is on a dryer, it is heated up and the water inside the sheet is transferred into felt. Assuming that D is the equivalent diffusion coefficient of all mass transfer mechanisms, Fick's second law can be applied

$$\frac{\partial M_f(x,t)}{\partial t} = D\frac{\partial^2 M_f(x,t)}{\partial x^2} \tag{2.101}$$

where M_f is the felt moisture content, x is the abscissa in felt thickness direction, and t is time.

It is assumed that the initial moisture content of the felt is uniform

$$M_f(x,t)|_{t=0} = M_{f0} \tag{2.102}$$

and the moisture content at the surface contacting the paper sheet is the same as that of the wet sheet

$$M_f(x,t)|_{x=0} = Ms_0 e^{-\alpha t} \tag{2.103}$$

where α simulates the varying sheet moisture content on dryer cylinder. Since the contact time of sheet with felt in a dryer is very short and the capacity of felt is much greater than that of the sheet, it is reasonable to assumed that the moisture content of the felt on the other surface remains constant

$$M_f(x,t)|_{x=H} = M_{f0} \tag{2.104}$$

where H is the thickness of the felt.

With the initial condition (2.102) and boundary conditions (2.103) and (2.104), the solution of the partial differential equation (2.101) is obtained by applying the suitable transformation (Carslaw and Jaeger, 1959), so that

$$
\begin{aligned}
M_f(x,t) \; = \; & M_{f0} + 2DH \sum_{n=1}^{\infty} (M_{s0}e^{-\alpha t}\frac{1}{Dn^2\pi^2 - \alpha H^2} \\
& - (\frac{M_{s0}}{Dn^2\pi^2 - \alpha H^2} - \frac{M_{f0}}{Dn^2\pi^2})e^{-Dn^2\pi^2 t/H^2} \\
& - \frac{M_{f0}}{Dn^2\pi^2})sin\frac{n\pi x}{H}
\end{aligned}
\tag{2.105}
$$

The mass transfer rate through the surface can be described by

$$
\begin{aligned}
N_A \; = \; & -D\frac{\partial M_f}{\partial x}\Big|_{x=0} \\
= \; & 2\pi D^2 \sum_{n=1}^{\infty} (\frac{M_{s0}}{Dn^2\pi^2 - \alpha H^2}e^{-\alpha t} - (\frac{M_{s0}}{Dn^2\pi^2 - \alpha H^2} \\
& - \frac{M_{f0}}{Dn^2\pi^2})e^{-Dn^2\pi^2 t/H^2} - \frac{M_{f0}}{Dn^2\pi^2})n
\end{aligned}
\tag{2.106}
$$

where N_A is the molar rate of moisture diffusion per unit area. The mass rate of evaporation g is thus

$$
g = MN_A
\tag{2.107}
$$

where M is the molecular weight of water.

In practical modeling, the first three terms in equation (2.106) are accurate enough to represent the mass transfer rate.

The mass transfer rate in the above equations is represented by the moisture content. It can be represented by the vapour pressure by applying

$$
M_{s0} = \frac{kP_{w0}}{T} \qquad M_{f0} = \frac{kP_a}{T_a}
\tag{2.108}
$$

where P_{w0} and P_a are the vapour pressure on the sheet surface and the partial vapour pressure in the surrounding air, T and T_a are the sheet temperature and the surrounding air temperature, k is constant coefficient.

The constant coefficients α, D and k are determined from the static moisture content into and out of each dryer cylinder.

Sheet in the draw ($\widehat{BC}$)

The mass and heat transfer in the draw is a double side free evaporation process. The mass and heat transfer rates can be described by

$$g = \frac{2KM}{RT}(P_w - P_a) \tag{2.109}$$

$$q = 0.664(\lambda \frac{P_r^{\frac{1}{3}}}{\nu^{\frac{1}{2}}})(T - T_a)/(\frac{l_e}{S_p}) \tag{2.110}$$

with

$$K = 0.664(\frac{S_p}{l_e})^{\frac{1}{2}}(\frac{D'}{\nu^{\frac{1}{2}}}S_c^{\frac{1}{3}}) \tag{2.111}$$

where g and q are the heat and mass transfer rates of the sheet, S_c, P_r, λ and ν are the Schmidt number, Prandtl number, thermal conductivity and dynamic viscosity of the surrounding air, respectively. D' is the diffusivity of water molecule in the surrounding air. S_p and l_e are machine speed and equivalent distance, respectively.

Lumped model

The model relating the sheet moisture content on reel to the surface temperature of the dryer cylinders can be obtained by applying mass and thermal balance principles using equations (2.100), (2.106)-(2.111), as described in the following.

At the abscissa y in the machine direction, we consider a sheet element of length Δy, width L and moving at the speed S_p (Figure 2.10). The mass balance gives

$$(Ms(y + \Delta y) - Ms(y))BwL\Delta y = -gL\Delta y\Delta t \tag{2.112}$$

and the thermal balance gives

$$(T(y + \Delta y) - T(y))Bw(C_f + Ms(y)C_w)L\Delta y$$
$$= ((T(y) - 273)g - q - gH_s)L\Delta y\Delta t \tag{2.113}$$

where $\Delta t = \Delta y/S_p$, Bw is the dry basis weight of the sheet, H_s is the enthalpy of water vapour at T=273°K, C_f and C_w are the specific heat of the fibre and water, respectively. Rearranging equations (2.112) and (2.113) gives

$$Ms(y + \Delta y) = Ms(y) - \frac{g}{Bw}\Delta t \tag{2.114}$$

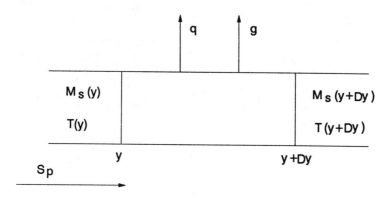

Figure 2.10: Paper sheet element in drying section

$$T(y + \Delta y) = T(y) + \frac{(T(y) - 273)g - q - gH_s}{Bw(C_f + Ms(y)C_w)}\Delta t \qquad (2.115)$$

The iterative computation of equations (2.114)-(2.115) on the dryer and then in the draw is successively executed for all the dryer cylinders, from the beginning to the end of the drying section. It must, of course, take into account the capacity of the felt. This iterative computation is quite easily performed by means of a digital computer and it leads to a central moisture content model

$$
\begin{aligned}
Ms(k+1) \;=\; & 0.9167Ms(k) - \sum_{i=1}^{7} a_{mi}T_{di} + 0.01758Bw(k) \\
& +\; 0.0169Ms_1(k) + 0.1524S_p(k) \\
& -\; 0.00261T_a(k) + 0.00451P_a(k) \qquad (2.116)
\end{aligned}
$$

with

$$
\begin{aligned}
(a_{m1}, a_{m2}, \cdots, a_{m7}) \;=\; & (0.02711, 0.03555, 0.04058, 0.05767 \\
& 0.05445, 0.05968, 0.06139)
\end{aligned}
$$

where T_{di} is the surface temperature of the ith sheet dryer cylinder , Ms and Ms_1 are the sheet moisture contents at the entrance and the end of the drying section, respectively.

2.5.3 Dryer surface temperature model

The dryers are heated up by steam, and transfer heat to sheet and surrounding air mainly through conduction and convection. The steam supply

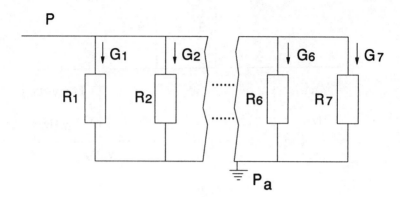

Figure 2.11: Simulated electricity network for steam supply system

system in the drying section can be simulated by an analogous electrical network. Figure 2.11 is the simulated electricity network for the steam supply system.

Let us consider the ith dryer. If its thermal capacitance is M_{ci}, and the thermal energy entering and leaving the dryer is Q_{ini} and Q_{outi}, respectively. The thermal balance gives

$$M_{ci}\frac{dT_{di}}{dt} = Q_{ini} - Q_{outi} \qquad (2.117)$$

The thermal energy entering the dryer is supplied by the steam flowing into it

$$Q_{ini} = G_i(H_s - H_w) \qquad (2.118)$$

where G_i is the flow rate of steam, H_s and H_w are the enthalpy of the steam and the condensate, respectively.

The thermal energy flowing out of the dryer includes the energy transferred to sheet and to the ambient medium. Experiments have shown that the latter is proportional to the former. Using the subscript "in" and "out" to represent the variables at the entry and exit of the dryer, respectively, we have

$$
\begin{aligned}
Q_{outi} = {} & \beta((C_f + M_{sini}C_w)(T_{outi} - T_{ini}) \\
& + (M_{sini} - M_{souti})(H_e - C_wT_{outi}))BwS_pL \qquad (2.119)
\end{aligned}
$$

where H_e is the enthalpy of the vapour at temperature T_{outi}, and β is set to be 1.23.

In static operating conditions, the right side of equation (2.117) is equal to zero, thus

$$G_i = \frac{Q_{outi}}{H_s - H_w} \qquad (2.120)$$

From Figure 2.11, the relationship between the steam pressure and the steam flow rate is obtained

$$G_i = \frac{P - P_a}{R_i} \qquad (2.121)$$

where R_i is the flow resistance, and P_a is pressure inside the dryer. It is assumed that the pressure inside the dryer is the same as the surrounding air pressure. The flow resistance can be calculated from equation (2.121) by using the steady state value of steam flow rate and steam pressure.

From equations (2.117)-(2.121), the dynamic model of the dryer surface temperature is obtained in the form of

$$
\begin{aligned}
T_{di}(k+1) &= a_{di1}T_{di}(k) + a_{di2}P(k) + a_{di3}M_{sini}(k) + a_{di4}Bw(k) \\
&+ a_{di5}S_p(k) + a_{di6}T_s(k) + a_{di7}T_a(k) + a_{di8}P_a(k) \qquad (2.122)
\end{aligned}
$$

where i=1,2, $\cdots$, 7; T_s is steam temperature.

2.5.4 Equivalent dryer model

The drying section consists of 7 sheet dryer cylinders. All the dryers affect the reel moisture content. If every dryer surface temperature serves as one state in state space model, the model will be of high dimension. To obtain an accurate model of low dimension, a parameter which reflects the dynamics of whole drying section is required. For this reason, we introduce the concept of "equivalent dryer", whose temperature is a weighted summation of the surface temperatures of all dryers. The weights in the summation are determined by the effects of the dryer surface temperatures on the moisture content.

From equation (2.116), we define the "equivalent dryer" temperature as

$$
\begin{aligned}
T_d(k) &= \sum_{i=1}^{7} \frac{a_{mi}}{\sum_{j=1}^{7} a_{mj}} T_{di}(k) \\
&= \sum_{i=1}^{7} a'_{mi} T_{di}(k) \qquad (2.123)
\end{aligned}
$$

with

$$
\begin{aligned}
(a'_{m1}, a'_{m2}, \cdots, a'_{m7}) &= (0.0809, 0.1061, 0.1211, 0.1721 \\
&\quad 0.1625, 0.1781, 0.1832)
\end{aligned}
$$

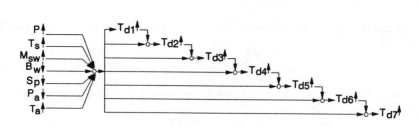

Figure 2.12: Relationship among process variables

The moisture content model (2.116) can be rewritten by using the "equivalent dryer" temperature T_d

$$
\begin{aligned}
Ms(k+1) &= 0.9167Ms(k) - 0.3351T_d(k) + 0.01698Ms_1(k) \\
&+ 0.01758Bw(k) + 0.1524S_p(k) \\
&- 0.00261T_a(k) + 0.00451P_a(k)
\end{aligned} \tag{2.124}
$$

Since all the cylinder surface temperatures in the drying section the contribute to "equivalent dryer" temperature, the varying of dryer temperature distribution will not affect the validity of the "equivalent dryer" model. The problem is how to analyze the dynamics of the "equivalent dryer". Figure 2.12 shows the relationship of the dryer surface temperatures to the influence variables. The marks "↑" and "↓" represent the increase and decrease of the variables.

The dynamics of the "equivalent dryer" can be obtained by using the following procedure according to Figure 2.12:

(1) changing the steam pressure by step ΔP and calculating the response of T_{d1} and Ms_1, and then T_{d2} - T_{d7} using equations (2.116) and (2.122);

(2) calculating the response of "equivalent dryer" temperature $\Delta T_d(t)$ and the final change $\Delta T_d(\infty)$. The time when the response reaches 63.2% of the final change is taken as the time constant. The model gain is $\Delta T_d(\infty)/\Delta P$;

(3) using the same method to calculate the gain of T_d to other variables.

The dynamics of the "equivalent dryer" then is described as

$$
\begin{aligned}
T_d(k+1) &= 0.9892T_d(k) + 0.1244P(k) - 2.6746 \times 10^{-4}Bw(k) \\
&- 3.2443 \times 10^{-4}Ms_1(k) - 0.00432S_p(k) \\
&+ 4.971 \times 10^{-4}T_s(k) + 2.786 \times 10^{-5}T_a(k) \\
&- 1.065 \times 10^{-5}P_a(k)
\end{aligned} \tag{2.125}
$$

Equations (2.124) and (2.125) give the dynamic model of the whole drying section.

2.5.5 Basis weight model

The model for the dry basis weight, which is invariant in the drying section, has been developed in the wire and press section. It keeps constant in the drying section. The model for the basis weight on the reel can be obtained by combining the dry basis weight model and the moisture content model.

Obviously, we have

$$Bw_1 = Bw(1 - Ms) \tag{2.126}$$

where Bw is the reel basis weight, Bw_1 is the dry basis weight after the second press. Taking deviation on both sides of the above equation gives

$$\Delta Bw_1 = \Delta Bw - Ms_0 \Delta Bw - Bw_0 \Delta Ms \tag{2.127}$$

or

$$\Delta Bw = \frac{1}{1 - Ms_0} \Delta Bw_1 + \frac{Bw_0}{1 - Ms_0} \Delta Ms \tag{2.128}$$

Substituting the steady state values to the equation and considering the measurement range of the sensors, we have

$$Bw = 1.089 Bw_1 + 0.2614 Ms \tag{2.129}$$

Considering the time delay between the basis weight on the reel and the basis weight after the second press, we have

$$Bw(k + 1) = 1.089 Bw_1(k - 1) + 0.2614 Ms(k) \tag{2.130}$$

2.6 Model Accuracy Test and Conclusions

So far, the models of all the three sections have been developed. The global model for the paper machine can be easily obtained by interconnecting these models.

A number of experiments have been carried out on a condenser paper machine to evaluate the model. Figures 2.13 and 2.14 show the output of the model and the actual process when the thick stock control aperture has a step change. Figure 2.15 compares the actual moisture content and the model output when the opening of the steam control valve is changed. It can be seen that the model output fits the actual basis weight and moisture content data quite well, which implies that the model is accurate.

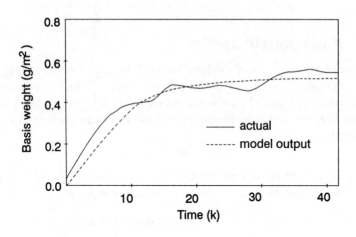

Figure 2.13: Basis weight to the step change of pulp valve

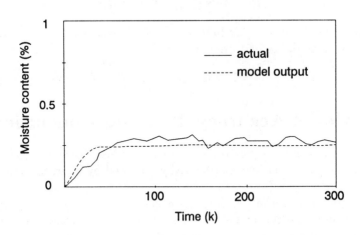

Figure 2.14: Moisture content to the step change of pulp valve

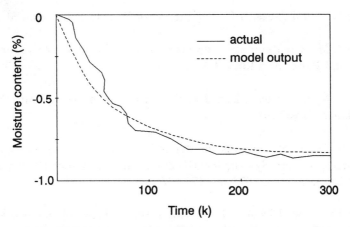

Figure 2.15: Moisture content to the step change of steam valve

In this chapter, a modeling technique was proposed for a super-thin condenser paper machine and a pressurized headbox. The modeling was mainly based on process mechanism analysis. However, it incorporated on-site tests and experiments as well as block diagram analysis. A new concept, namely "equivalent dryer", was introduced to reduce the dimension of the system model and to improve the accuracy of the model. The steam supply system was simulated by an electricity network. A number of assumptions were introduced to simplify the modeling and make the resulting model workable. This modeling technique is easy to use and can be applied to different kind of paper machines. Experiments have shown that the models have satisfactory accuracy.

2.7 References

Anderson L., and Back E.L. The effect of temperature up to 90°C on dewatering of paper webs, Evaluate in a press simulator. Proc of TAPPI 1981 Engineering conference, 1981, pp 311-323

Åström K.J. Computer Control of a paper machine–an application of linear stochastic control theory. IBM J. Rev. Dev. 1967; 11:389-405

Baines W.D. Analysis of transients effects in drying of paper. Pulp and Paper Magazine of Canada 1973; 74:T34-T40

Carslaw H.S. and Jaeger J.C. Conduction of Heat in Solid. Oxford, 1959

Codieux S.M. A mathematical drainage method for fourdrier paper machine. J. Pulp and Paper Sci. 1983; TK:111-116

Conlson J.M. and Richardson J.F. Chemical Engineering, Third Edn. Vol. 1, Pergamon Press, 1977

Gentile S. Different methods for dynamic identification of an experimental paper machine. In: System Identification and Parameter Estimation. New York, Wiley and Sons, 1974, pp 473-483

Hartley F.T. and Richards R. J. Hot surface drying of paper-the development of a diffusion mode. TAPPI 1974; 57(3):157-160

Knight R.L. and Kirk L.A. Simulation of the paper machine drying section. Presented at International Water Removal Symposium, British Paper and Board Industries Federation, London, 1975

Kugushev K.S. Theory of paper-making and water removal processes. (in Chinese, originally translated from Russian) Light Industry Press Ltd. Beijing, 1982

Lemaitre J. Veyre, Lebeau B. and Foulard C. Method for systematic analysis of paper machine multicylinder drying section. Proc of the 4th IFAC Conf.on Instrumentation and Automation in the Pulp, Plastics and Polynomication Industries, Gand, Belgium, 1980, pp 261-270

Xia Q. and Sun Y. Computer control of basis weight and moisture content for a condenser paper machine. Chinese Pulp and Paper 1989; 3:73-81

Xia Q., Rao M., Shen X., Ying Y. and Zurcher J. Systematic modeling and decoupling control of a pressurized headbox. Proc. 1993 Canadian Conf. on Electrical and Computer Engineering, Vancouver, 1993, pp 962-965

Xia Q., Rao M., Sun Y. and Ying Y. Modeling and control for drying section of paper machines. Proc. Control System'92, Whistler, B.C., Canada, 1992, pp 7-11

CHAPTER 3
ROBUST CONTROL

A pressurized headbox with significant uncertainty in dynamics is considered and assumed to be represented by a set of multiple finite-dimensional linear time-invariant models. The principal purpose is to design a fixed robust controller in a unit output feedback configuration such that the resulting feedback system simultaneously satisfies the given performance objectives for all models in the set. A new approach is developed in frequency domain and its effectiveness is illustrated by simulations.

The problem is defined in Section 3.2.1. An approach to the construction of its solutions is outlined, which includes two main steps as the achievement of simultaneously diagonal dominance, and the realization. The former is presented in Section 3.2.2 and the later is discussed in Section 3.2.3. Section 3.2.4 deals with a design procedure and its application. Some conclusions are finally given in Section 3.3.

3.1   Introduction

A pressurized headbox is a typical nonlinear multivariable system (Xia, et al, 1993). Its operating conditions must be adjusted to various product grades and different production rates. It is impossible to represent its dynamics by a single linear model. The dynamics of a pressurized headbox can be better represented by a set of linear models. In this chapter, the design of a robust controller for a multivariable system specified by a set of p finite-dimensional linear time-invariant models is discussed. A physical realizable, finite-dimensional linear time-invariant controller in a unit output configuration is designed such that the resulting feedback system simultaneously satisfies performance specifications for all models in the set. The performance

53

specifications considered here include stability, regulation, tracking, interaction and transient behaviour. A new approach to such a controller design is developed in the frequency domain setting. All computations involved in the construction of the controller contain only real matrix operations and can be carried out by well-developed algorithms and software of matrix computations and linear programming. The applicability and effectiveness of this approach is illustrated by the design and simulation of a pressurized headbox system.

3.2 Multi-Model Robust Control

3.2.1 Problem formulation and simplification

Consider a multivariable discrete-time system with p $m \times r$ proper rational impulse transfer function models $G^1(z)$, $G^2(z)$, . . ., $G^p(z)$ given. The following control law is introduced,

$$U(z) = G_c(z)E(z) \tag{3.1}$$

where U and E are control and error vectors, respectively, as shown in Figure 3.1, and G_c must be proper for physical realizability, i.e., $\lim_{z \to \infty} G_c(z) < \infty$. The problem at hand (Problem 1) is to determine a single controller G_c such that the resulting feedback system satisfies the following five performance objectives for all models G^l, $(l = 1, 2, ..., p)$.

(a) Stability: The closed-loop characteristic polynomial (CLCP) is a stable polynomial, i.e., all its roots are inside the open unit circle in Z-plane. This circle is denoted by R_1.

(b) Regulation: With $R(z) = 0$, output vector Y decays to zero at the steady-state for any D in the specific class of disturbance vectors.

(c) Tracking: With $D(z) = 0$, error vector E decays to zero at the steady-state for any R in the specific class of reference vectors.

(d)Weak interaction among different loops.

(e) Desirable transient response characteristics: It may include percent overshoot, rise time and settling time.

It should be pointed out that these performance specifications are often required in engineering applications.

The reference vector R and disturbance vector D are assumed to be generated, respectively, from $M_r^{-1}(z)$ and $M_d^{-1}(z)$. Let $\phi_r(z)$ and $\phi_d(z)$ be their monic minimal polynomials respectively. Without loss of generality, ϕ_r and ϕ_d

can be assumed to be completely unstable polynomials, i.e., none of their roots is in R_1. The least common factor of ϕ_r and ϕ_d is denoted by ϕ. It then follows from the robust servomechanism theory (Callier and Desoer, 1982) that the regulation and tracking with stability is solvable for all G^l if and only if

(1) $r \leq m$

(2) None of G^l, ($l=1,2,..., p$), has the same zeros as those of ϕ.

It is assumed that the conditions (1) and (2) hold true, thus the robust regulation and tracking with stability are solvable, and, further, the controller for this purpose must have the following form:

$$G_c = G_{co}\bar{G}_c \tag{3.2}$$

where $G_{co} = \phi^{-1}I$ and $\bar{G}_c$ simultaneously stabilizes all $\bar{G}^i = G^l G_{co}$, ($l=1, 2,..., p$). With such a G_c, the performance specifications (a) through (c) have been satisfied. For objective (d), it is here characterized by diagonal dominance (Rosenbrock, 1974). Let $\bar{G}_c = G_{c1}G_{c2}$. G_{c1} is used to make $\bar{G}^l G_{c1}$ diagonal dominance for all i while G_{c2} simultaneously stabilizes them.

Now objective (e) remains to be considered. For this purpose, it is assumed that the desirable transient response characteristics are satisfied if every one of the p CLCPs (each for one model) has its roots all in subregion R of the stability region R_1. A closed-loop system with all roots of its CLCP in R is said to be nicely stable (Ackermann, 1980). Thus, specification (e) will also be satisfied if G_{c2} simultaneously nicely stabilizes all $\bar{G}^l G_{c1}$, ($l=1, 2,..., p$).

In view of these observations that Problem 1 is largely simplified into the following one (Problem 2): Let a system be described by its p models G^1, $G^2,..., G^p$, and also let a completely unstable polynomial ϕ defined above be given. Denoting $G^l \phi^{-1}$ by $\bar{G}^l$, then the problem is to determine a controller $\bar{G}_c = G_{c1}G_{c2}$ with $\phi^{-1}\bar{G}_c = G_c$ being proper such that G_{c1} simultaneously makes all $\bar{G}^l G_{c1}$, ($l=1, 2,..., p$), diagonal dominance, and G_{c2} simultaneously achieves nice stabilization of all $\bar{G}^l G_{c1}$, ($l=1, 2, ...,p$).

3.2.2   Achieving simultaneous dominance

For a single model, various methods of achieving or increasing diagonal dominance are variable, which include elementary operations and pseudo-diagonalization, etc. Most of them can be extended such that they are applicable to the case of multiple models. In this section, the pseudo-diagonalization of a single model is extended to the simultaneous pseudo-diagonalization of multiple models.

Let $\bar{G}^1$, $\bar{G}^2,...,\bar{G}^p$ of dimension $m \times r$ be the given process dynamics, G_{c1} of dimension $r \times m$ be a precompensator, and $\mathcal{D}$ be a specific contour. The purpose is to determine G_{c1} with its structure as simple as possible such that $\bar{G}^l G_{c1}$, $(l = 1, 2, ..., p)$, are all diagonally dominant on $\mathcal{D}$.

Now, a constant precompensator is first considered, i.e., $G_{c1} = K$, a constant matrix. Choosing some $z = c$ on $\mathcal{D}$ and considering column k of $\bar{G}^l K$, $(l = 1, 2, ..., p)$, the elements q_{jk} of this column are

$$q_{jk}(c) = \sum_{u=1}^{r} \bar{g}_{ju}^l(c) k_{uk} = \sum_{u=1}^{r} (\alpha_{ju}^l + i\beta_{ju}^l) k_{uk} \tag{3.3}$$

where we have written

$$\bar{g}_{ju}^l = \alpha_{ju}^l + i\beta_{ju}^l$$

Now let us choose k_{1k}, k_{2k}, ..., k_{rk} so that

$$\sum_{l=1}^{p} \sum_{j=1, j\neq k}^{m} |q_{jk}(c)|^2 \tag{3.4}$$

is as small as possible subject to the constraint

$$\sum_{u=1}^{r} k_{uk}^2 = 1 \tag{3.5}$$

Using a Lagrange multiplier λ, we need to minimize the following cost function

$$
\begin{aligned}
J_k &= \sum_{l=1}^{p} \sum_{j=1, j\neq k}^{m} \left| \sum_{u=1}^{r} (\alpha_{ju}^l + i\beta_{ju}^l) k_{uk} \right|^2 + \lambda(1 - \sum_{u=1}^{r} k_{uk}^2) \\
&= \sum_{l=1}^{p} \sum_{j=1, j\neq k}^{m} \left\{ (\sum_{u=1}^{r} \alpha_{ju}^l k_{uk})^2 + (\sum_{u=1}^{r} \beta_{ju}^l k_{uk})^2 \right\} \\
&+ \lambda(1 - \sum_{u=1}^{r} k_{uk}^2)
\end{aligned}
\tag{3.6}
$$

Taking the partial derivative with respect to k_{vk}, we obtain

$$
\begin{aligned}
\frac{\partial J_k}{\partial k_{vk}} &= \sum_{l=1}^{p} \sum_{j=1, j\neq k}^{m} \left\{ 2(\sum_{u=1}^{r} \alpha_{ju}^l k_{uk}) \alpha_{jv}^l + 2(\sum_{u=1}^{r} \beta_{ju}^l k_{uk})^2 \beta_{jv}^l \right\} \\
&- \lambda 2 k_{vk} = 0 \qquad v = 1, 2, ..., r.
\end{aligned}
\tag{3.7}
$$

It can be rewritten as

$$\sum_{u=1}^{r} \left\{ \sum_{l=1}^{p} \sum_{j=1, j\neq k}^{m} (\alpha_{jv}^l \alpha_{ju}^l + \beta_{jv}^l \beta_{ju}^l) \right\} k_{uk} = \lambda k_{vk} \quad v = 1, 2, ..., r. \tag{3.8}$$

We now introduce the real symmetric matrix

$$A_k = (a_{vu}^{(k)}) = (\sum_{l=1}^{p} \sum_{j=1,j\neq k}^{m} (\alpha_{jv}^l \alpha_{ju}^l + \beta_{jv}^l \beta_{ju}^l)) \tag{3.9}$$

which is required to be at least positive semi-definite, so that its eigenvalues are real and non-negative. If we also introduce the column vector

$$k_k = (k_{uk})$$

Equation (3.8) can be organized as the form

$$A_k k_k = \lambda k_k \tag{3.10}$$

which is a standard eigenvector problem. If the eigenvector of A_k make (3.8) true, we can obtain

$$\sum_{l=1}^{p} \sum_{j=1,j\neq k}^{m} | q_{jk} |^2 = k_k^T A_k k_k = \lambda k_k^T k_k = \lambda$$

In order to minimize (3.4), we have to choose k_k corresponding to the smallest eigenvalue of A_k.

The eigenvector problem (3.10) is not difficult to solve because A_k is real and symmetric. Its solution gives us the k-th column in K, of unit length, which minimizes the sum of squares, of the magnitudes of the off-diagonal elements of column k of $\bar{G}^1 K$, $\bar{G}^2 K$,..., $\bar{G}^p K$.

We may solve a similar problem for each column of K. In this way we can obtain candidates for K which may satisfy our requirements. The procedure described above can be generalized in several ways. First, instead of minimizing (3.4), we may minimize the more general form

$$\sum_{t=1}^{N}(\sum_{l=1}^{p} \sum_{j=1,j\neq k}^{m} | q_{jk}(c_t) |^2 r_t) \tag{3.11}$$

subject to (3.5). In (3.11), r_t are real and positive, and the quantity to be minimized is a weighted sum of squares, at frequencies ω_1, ω_2, ..., ω_N, of the off-diagonal elements in column k of $\bar{G}^l k$, ($l=1,2,...,p$). An analysis similar to that given before shows that column k of K is given again by an eigenvector problem

$$B_k k_k = \lambda k_k \tag{3.12}$$

where

$$B_k = (b_{vu}^{(k)}) = \sum_{t=1}^{N} r_t \sum_{l=1}^{p} \sum_{j=1,j\neq k}^{m} (\alpha_{vjt}^l \alpha_{ujt}^l + \beta_{vjt}^l \beta_{ujt}^l)$$

α^l_{ujt} and β^l_{ujt} are defined as

$$\bar{g}^l_{uj}(c_t) = \alpha^l_{ujt} + i\beta^l_{ujt}$$

Similarly, choose the eigenvector k_k corresponding to the smallest eigen-value of B_k. The same set $\omega_1, \omega_2, ..., \omega_N$ may be used for each column of K, or different sets may be used for different columns.

A second generalization is to consider, with a more general form

$$G_{c1} = K_0 + K_1 z$$

If we write the k-th column of $G_{c1}(c)$ in the form

$$k_k^{(0)} + ic'' k_k^{(1)}$$

where $c = c' + ic''$, $k_k^{(0)} = k_k^0 + c' k_k^1$, $k_k^{(1)} = k_k^1$, k_k^0 and k_k^1 are the k-th column of k_0 and k_1 respectively. We can choose $k_k^{(0)}$ and $k_k^{(1)}$ to minimize (3.4) with constraint as

$$\sum_{j=1}^{r} \{(k_{jk}^{(0)})^2 + (k_{jk}^{(1)})^2\} = 1$$

This leads to another eigenvector problem

$$D_k \begin{pmatrix} k_k^{(0)} \\ k_k^{(1)} \end{pmatrix} = \lambda \begin{pmatrix} k_k^{(0)} \\ k_k^{(1)} \end{pmatrix}$$

where

$$D_k = \begin{pmatrix} A_k & c''C_k \\ c''C_k & (c'')^2 A_k \end{pmatrix}$$

where A_k given by (3.9) and C_k by

$$C_k = (c_{vu}^{(k)}) = \sum_{l=1}^{p} \sum_{j=1, j\neq k}^{m} (\alpha^l_{vj}\alpha^l_{uj} - \beta^l_{vj}\beta^l_{uj})$$

3.2.3   Simultaneous stabilization

When the simultaneously diagonal dominance has been achieved, diagonal matrix $G_{c2}=\text{diag}(g_{c21}, g_{c22}, ..., g_{c2m})$ can be used to implement single-loop compensators as required to realize simultaneously nice stabilization. Let the i-th diagonal element of $Q^l=\bar{G}^l G_{c1}$ be q^l_{ii}, and $P^+(q^l_{ii})$ be the number of those poles which are not in R. Then the following theorem can be established.

Theorem 1 *Let* $(I + Q^l G_{c2})$, $l=1,2,...,p$, *be diagonally dominant on* ∂R, *the boundary of* R, *and let none of their diagonal elements has any pole on* ∂R. *If for each* i $(i = 1, 2, ..., m)$, g_{c2i} *simultaneously nicely stabilizes all* q_{ii}^l, $(l = 1, 2, ..., p)$, *and*

$$P^+(\bar{G}^l) + P^+(G_{c1}) = \sum_{i=1}^{m} P^+(q_{ii}) \quad l = 1, 2, ..., p \tag{3.13}$$

then the p *closed-loop systems (each for one model) are all nicely stable.*

Proof: Consider arbitrary but fixed l, and let $q_{ii}^l g_{c2i}$ (resp. $det(I + QG_{c2})$) map ∂R into Γ_i (resp. Γ) which encircles the origin N_i (resp. N) times, $(i = 1, 2, ..., m)$, then the closed-loop system is nicely stable (Rosenbrock, 1974) if

$$N = -P_o^+ \tag{3.14}$$

where P_o^+ is number of open-loop poles which are not in R. P_o^+ is given by

$$\begin{aligned} P_o^+ &= P^+(G^l) + P^+(G_{c1}) + P^+(G_{c2}) \\ &= P^+(G^l) + P^+(G_{c1}) + \sum_{i=1}^{m} P^+(g_{c2i}) \end{aligned} \tag{3.15}$$

On the other hand, since $(I + Q^l G_{c2})$ is diagonally dominant on ∂R, by Rosenbrock's theorem (1974), we obtain

$$N = \sum_{i=1}^{m} N_i \tag{3.16}$$

Furthermore, since g_{c2i} nicely stabilizes q_{ii} for all i, it follows

$$N_i = -P_{oi}^+ = -(P^+(q_{ii}) + P^+(g_{c2i})) \tag{3.17}$$

where P_{oi}^+ is the number of the open-loop poles of the ith loop which are not in R. Thus, we have

$$\begin{aligned} N &= -\sum_{i=1}^{m}(P^+(q_{ii}) + P^+(g_{c2i})) \\ &= -\sum_{i=1}^{m} P^+(q_{ii}) - \sum_{i=1}^{m} P^+(g_{c2i}) \end{aligned} \tag{3.18}$$

From (3.13), (3.15) and (3.18), it follows that (3.14) holds for the l-th closed-loop system. But since i is arbitrary, the above statement applies to all *i*. The theorem is thus proven.   Q.E.D.

Following the theorem, we next consider the simultaneously nice stabilization of multiple SISO models. Unfortunately, it is found there exists noeffective method to directly construct a simultaneous stabilizing controller. Instead, we will here present two methods which give candidates for such controllers. A main advantage of our methods is much simple in computation.

Let an SISO plant has p strictly rational transfer function models q^1, q^2, ..., q^p, and let the controller to be determined by g. q^l and g are represented by the polynomial fraction forms

$$q^l = \frac{b^l}{a^l}, \quad g = \frac{y}{x} \tag{3.19}$$

For a fixed l, the CLCP is given as

$$h^l = a^l x + b^l y \tag{3.20}$$

(a^l, b^l), (x, y) and h^l are rewritten in the power forms

$$[a^l, b^l] = \sum_{i=1}^{n_l} [a_i^l, b_i^l] z^{n_l - i} \tag{3.21}$$

$$[x, y] = \sum_{i=0}^{N_c} [x_i, y_i] z^{n_c - i} \tag{3.22}$$

$$h^l = \sum_{i=0}^{n} p_i s^{n-i} \quad n = n_l + n_c \tag{3.23}$$

Substituting (3.21)-(3.23) into (3.20) and comparing the coefficients on both sides yields

$$S^l K = h^l \tag{3.24}$$

where

$$K^T = [x_0 \ y_0 \ x_1 \ y_1 ... x_{n_c} \ y_{n_c}]$$
$$(h^l)^T = [h_0 \ h_1 ... h_n]$$

$$S^l = S(a^l, b^l, n_c) = \begin{pmatrix} a_0 & b_0 & 0 & 0 & ... & 0 & 0 \\ a_1 & b_1 & a_0 & b_0 & ... & 0 & 0 \\ . & . & a_1 & b_1 & ... & 0 & 0 \\ . & . & . & . & . & . & . \\ a_{nl} & b_{nl} & . & . & ... & 0 & 0 \\ 0 & 0 & a_{nl} & b_{nl} & ... & a_0 & b_0 \\ . & . & . & . & ... & . & . \\ 0 & 0 & 0 & 0 & 0 & a_{nl} & b_{nl} \end{pmatrix}$$

which is a $(n_l + n_c + 1) \times 2(n_c + 1)$ constant matrix, and called as the Sylvester resultant. It is well known that S^l has full row rank if $n_c > n_l - 1$. If it is the case, (3.24) has a solution for any h^l obtained from a nice stable polynomial p^l.

Since (3.24) holds for all l, then combining such p equations yields

$$[S^1 \; S^2 \; ... \; S^p]K = [h^1 \; h^2 \; ... \; h^p] \qquad (3.25)$$

Using a result on the canonical parameter space of stable polynomials (Fam and Meditch, 1978) and its generalization to nicely stable polynomials (Ackermenn, 1980), we can establish a simultaneously nice stabilizing controller. Let R be an open circle with radius r_0 and center at c_0 on the real axis of the Z-plane. This circle intersects the real axis at $c_1 = c_0 - r_0$, and $c_2 = c_0 + r_0$. c_0 and r_0 are restricted to satisfy

$$max(|\, c_0 - r_0 \,|, |\, c_0 + r_0 \,|) < 1$$

It ensures that R is a proper subset of the stability region R_1. Let a polynomial of order n be given by

$$p(z) = z^n + b_1 z^{n-1} + ... + b_{n-1}z + b_n$$

A map M from an n-order real polynomial into an $(n+1)$-dimensional real vector h is defined as follows

$$h = M(p(z)) = [1 \; b_1 \; b_2 \; ... \; b_n]^T$$

Also let $(n+1)$ polynomials $p_i(z)$ be

$$p_i(z) = (z - c_1)^i (z - c_2)^{n-i} \quad i = 0, 1, ..., n$$

and the corresponding vector h_i be

$$h_i = M(p_i(z))$$

Then it has been shown (Fam and Meditch, 1978; Ackermann, 1980) that if a polynomial $p(z)$ is nicely stable (with respect to R) then there are $t_i \geq 0$, $(i = 0, 1, ..., n)$, such that

$$h = \sum_{i=0}^{n} t_i h_i$$

Denote

$$h = Ht \qquad (3.26)$$

where
$$H = [h_0 \ h_1 \ ... \ h_n], \quad t = [t_0 \ t_1 \ ... \ t_n]^T$$

This result is now applied to every h^l in (3.25), that is

$$S^l K = h^l \tag{3.27}$$

thus (3.25) becomes

$$\begin{pmatrix} S^1 \\ S^2 \\ \cdot \\ \cdot \\ \cdot \\ S^p \end{pmatrix} K = \begin{pmatrix} H & & & & \\ & H & & & \\ & & \cdot & & \\ & & & \cdot & \\ & & & & \cdot \\ & & & & H \end{pmatrix} \begin{pmatrix} t^1 \\ t^2 \\ \cdot \\ \cdot \\ \cdot \\ t^p \end{pmatrix} \tag{3.28}$$

where $t_i^l \geq 0$, $\sum_{i=1}^n t_i^l = 1$, $(l = 1, 2, ..., p)$. Let P be a nonsingular matrix such that
$$P[S^{1^T} \ S^{2^T} \ ... \ S^{p^T}]^T = [\tilde{S}^T \ 0]^T$$

where $\tilde{S}^T$ has full row rank. Partition P as

$$P = [P_1^T \ P_2^T]^T$$

where P_1 has the same number of rows as $\tilde{S}$. Premultiplying (3.28) by P gives

$$\tilde{S} K = P_1 \tilde{H} T \tag{3.29}$$

$$A T = 0 \tag{3.30}$$

$$t_i^l \geq 0 \quad \sum_{i=1}^n t_i^l = 1 \quad l = 1, 2, ..., p \tag{3.31}$$

where

$$\tilde{H} = \begin{pmatrix} H & & & & \\ & H & & & \\ & & \cdot & & \\ & & & \cdot & \\ & & & & \cdot \\ & & & & H \end{pmatrix}$$

$$T = [t^1 \ t^2 \ ... \ t^p]^T, \quad A = P_2 \tilde{H}$$

Since $\tilde{S}$ has full row rank, (3.29) always has a solution. Therefore the solvability of (3.28) is equivalent to the one of (3.30) and (3.31). We have reached an easy-solving form in which there is no unknown K. Furthermore, linear programming can now be used to determine whether equations (3.30) and (3.31) have a solution T, and to compute the solution when it exists (Bazaraa and Jarvis, 1977). Let $\tilde{T}$ be a vector with the dimension equal to the number of rows of $\tilde{A}$. Consider the following linear programming problem

$$LP: \quad J = min \sum_i t_i \qquad (3.32)$$

subject to $\tilde{A}T + \tilde{T} = \tilde{b}$, where

$$\tilde{A} = \begin{pmatrix} A & & & & & 0 \\ 1 \ 1 \ \dots \ 1 & & & & \\ & & 1 \ 1 \ \dots \ 1 & & \\ & & & \dots & \\ 0 & & & & 1 \ 1 \ \dots \ 1 \end{pmatrix}$$

$$\tilde{b}^T = [0 \ 1 \ 1 \ \dots \ 1]$$

By the theory of linear programming, (3.30) and (3.31) have a solution T if and only if J in LP is zero. If J is indeed zero, the candidate for a simultaneously nice stabilizing controller can be obtained from the solution of (3.29) with T being replaced by any optimal solution T^* of LP.

An alternative method of constructing a candidate is suggested for the stabilizing controller. Let the n closed-loop poles expected by the designer be $\lambda_1, \lambda_2, ..., \lambda_n$, which are all in R.

The corresponding expected closed-loop characteristic polynomial is

$$P_e(z) = (z - \lambda_1)(z - \lambda_2)...(z - \lambda_n) = z^n + P_1^e z^{n-1} + ... + P_n^e \qquad (3.33)$$

Its coefficient vector is

$$h^e = m(P_e(z)) = [1 \ P_1^e \ ... \ P_n^e]^T$$

In (3.25), we take all h^l, (l=1,2,...,p) as h^e. Since (3.25) generally has no solution, an exact solution is not recommended. We only need its least square solution. This implies that we construct such a controller which makes all p CLCPs approach the expected CLCP as close as possible.

Finally, it should be stated that the constructed controller from the methods mentioned above is not necessarily a simultaneously nice stabilizing controller, and therefore it is to be checked whether the resulting p CLCPs are all nice stable by some numerical approach.

3.2.4 Design procedure and application

Based on the development in the previous sections, a systematical design procedure for the construction of a controller which solves problem 1 is summed up as follows:

Given data: $p_{m \times r}$ rational transfer function matrix models G^1, G^2,..., G^p, a completely unstable polynomial ϕ and a nice stability region R.

Step 1 If $r < m$, Problem 1 has no solution; Otherwise, form $\bar{G}^l = G^l \phi^{-1}$, $l=1,2,...,p$.

Step 2 With the methods presented in in section 3.1.3, compute a precompensator G_{cl} such that $Q^l = \bar{G}^l G_{cl}$, $(l=1,2,...,p)$, are all diagonally dominant on ∂R.

Step 3 With the methods given in Section 3.1.4, compute $G_{c2}=\text{diag}(g_{c21}, g_{c22}, ..., g_{c2m})$ such that each g_{c2i} simultaneously nicely stabilizes q^l_{ii}, $(l=1,2,...p)$, and at the same time the resulting $(I+Q^L G_{c2})$, $(l=1,2,...,p)$, should be ensured to be diagonally dominant on ∂R.

Step 4 From the designed controller $G_c = \phi^{-1} G_{c1} G_{c2}$, the closed-loop simulations can be made if required.

In the remainder of this section, the controller design algorithm is applied to a pressurized headbox.

The description of the pressurized headbox has been given in Chapter 2. The outputs of the system are total head H and liquid level L, the outputs are fan pump speed u_1 and vacuum valve opening u_2. Considering that the system has nonlinear characteristics, we describe the headbox by three linear time-invariant models under different operating condition

$$Y(s) = G^l(s)U(s), \qquad (l = 1, 2, 3)$$

where $Y(s) = (H(s)\ L(s))^T$, $U(s) = (u_1(s)\ u_2(s))^T$. The transfer functions $G^l(s)$ $(l = 1, 2, 3)$ are given by

$$G^1(s) = \begin{pmatrix} \frac{1.75(3.895s+1)}{2.63s+1} & \frac{-0.15783s}{2.63s+1} \\ \frac{1.75(0.1788s+1)}{2.63s+1} & \frac{1.54}{2.63s+1} \end{pmatrix}$$

$$G^2(s) = \begin{pmatrix} \frac{1.78(4.215s+1)}{2.82s+1} & \frac{-0.1632s}{2.82s+1} \\ \frac{1.78(0.1876s+1)}{2.82s+1} & \frac{1.55}{2.82s+1} \end{pmatrix}$$

$$G^3(s) = \begin{pmatrix} \frac{1.80(4.613s+1)}{3.03s+1} & \frac{-0.1710s}{3.03s+1} \\ \frac{1.80(0.1988s+1)}{3.03s+1} & \frac{1.56}{3.03s+1} \end{pmatrix}$$

All the disturbances and references are assumed to be step signals. Therefore, in order to achieve asymptotic tracking and regulation, we may take

$$G_{c0} = \frac{1}{s}$$

Choosing $z = c = 0.3$ in the evaluation of equation (3.3) and using the method presented in Section 3.3.3, we obtain the precompensator G_{c1}

$$G_{c1} = \begin{pmatrix} 0.6368 & 0.0122 \\ -0.7711 & -0.9999 \end{pmatrix}$$

The equivalent transfer functions $Q^l = G^l \phi G_{c1}$ $(l = 1, 2, 3)$ are

$$Q^1(s) = \begin{pmatrix} \frac{4.4621s+1.1143}{2.63s^2+s} & \frac{0.0746s-0.0214}{2.63s^2+s} \\ \frac{0.1992s-0.0731}{2.63s^2+s} & \frac{-0.0038s-1.5612}{2.63s^2+s} \end{pmatrix}$$

$$Q^2(s) = \begin{pmatrix} \frac{4.9033s+1.1334}{2.82s^2+s} & \frac{0.0716s-0.0217}{2.82s^2+s} \\ \frac{0.2126s-0.0617}{2.82s^2+s} & \frac{-0.0041s-1.5716}{2.82s^2+s} \end{pmatrix}$$

$$Q^3(s) = \begin{pmatrix} \frac{5.3604s+1.1462}{3.03s^2+s} & \frac{0.0708s-0.0220}{3.03s^2+s} \\ \frac{0.2279s-0.0567}{2.63s^2+s} & \frac{-0.0044s-1.5819}{3.03s^2+s} \end{pmatrix}$$

Using the method presented in Section 3.2.3, the simultaneous stabilizing controller G_{c2} is obtained as follows

$$G_{c2}(s) = \begin{pmatrix} \frac{0.2905s+0.2472}{0.3545s+0.1785} & 0 \\ 0 & -\frac{0.6449s+0.2545}{0.3534s+0.6854} \end{pmatrix}$$

We thus obtained the designed controller for the pressurized headbox

$$G_c(s) = G_{c0}(s)G_{c1}(s)G_{c2}(s)$$

Figure 3.1 shows the block diagram of the control system. Figures 3.2 and 3.3 present the responses of total head and liquid level when the setpoint for total head is changed, and Figures 3.4 and 3.5 present that when the setpoint for liquid level changed. It is obvious from the figures that the performance specifications assigned have been achieved. The closed-loop system has satisfactory transient responses and the interactions between the two control loops are reduced to a very low level.

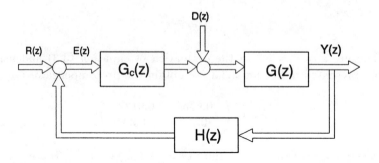

Figure 3.1: Block diagram of the multi-model robust control system

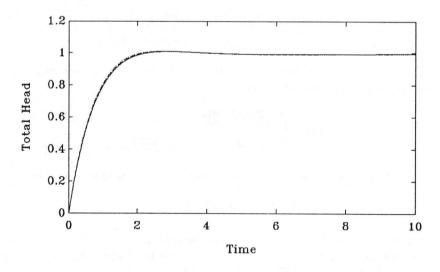

Figure 3.2: Response of total head with total head set-point unit step change

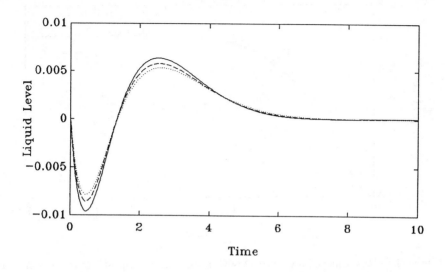

Figure 3.3: Response of liquid level with total head set-point unit step change

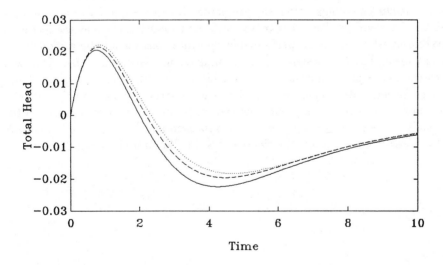

Figure 3.4: Response of total head with liquid level set-point unit step change

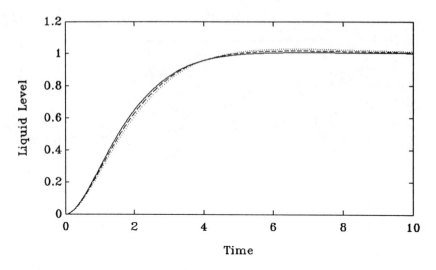

Figure 3.5: Response of liquid level with liquid level set-point unit step change

3.3   Conclusions

In this chapter, a new approach for designing a fixed robust controller in a unit feedback configuration was presented. It can be applied to a multivariable system with multiple linear time-invariant models such that the feedback system satisfies the given performance specifications for all these models. It was shown that the design can be completed in two main steps. One was determining a precompensator which achieves simultaneously diagonal dominance for multiple models, and another was constructing a diagonal controller of which each diagonal element simultaneously nicely stabilizes multiple SISO models of the corresponding loop. The approach had been used in the control of a pressurized headbox. Its effectiveness was illustrated by simulations.

3.4 References

Ackermann J. Parameter space design for robust control. IEEE Trans. Autom. Contr. 1980; AC-25:1058-1072

Bazaraa M.S. and Jarvis J.J. Linear programming and network flows. John Wiley and Sons, New York, 1977

Callier F.M. and Desoer C.A. Multivariable feedback systems. Springer-Verlag, New York, 1982

Chen M.J. and Desoer C.A. Necessary and sufficient conditions for robust stability of linear distributed feedback systems. Int. J. Control 1982; 35:255-267

Doyle J.C. and Stein G. Multivariable feedback design concepts for a classical/modern synthesis. IEEE Trans. Autom. Contr. 1981; AC-26:4-16

Doyle J.C. Analysis of feedback systems with structured uncertainties. IEE Proc. 1982; 129D:242-250

Fam A.T. and Meditch J.S. A canonical parameter space for linear systems design. IEEE Trans. Autom. Contr. 1978; AC-23:454-458

Garbow B.S. Matrix eigensystem routines EISPACK guide extension. Springer-Verlag, New York, 1977

Kokotovic P.V. Recent trends in feedback design: an overview. Automatica 1985; 21:225-236

Kwakernaak H. Minimax frequency domain performance and robustness optimization of linear feedback systems. IEEE Trans. Autom. Contr. 1985; AC-30:994-1004

Lehtomaki N.A. Robustnees and modeling error characterization. IEEE Trans. Autom. Contr. 1984; AC-29:212-220

Postlethwaite I. and Foo Y.K. Robustnees with simultaneous poles and movement across the jw-axie. Automatica 1985; 21:433-443

Rosenbrock H.H. Computer-aided control system design. Academic Press, London, 1974

Tzafestas S.G. (editor) Multivariable control: new concepts and tools. Holland, 1984

Xia Q., Rao M., Shen X., Ying Y. and Zurcher J. Systematic modelling and decoupling design of pressurized headbox. Proc Canadian conference on electical and computer engineering, Vancouver, 1993, pp 962-965

CHAPTER 4
PREDICTIVE CONTROL

The major difficulties in paper-making process control may arise from the following reasons: (1) some process states are unmeasurable; (2) there are long time delays; (3) there are significant parameter variations; (4) there are strong couplings between basis weight and moisture content control; (5) there are measurable and unmeasurable process disturbances. In this chapter, we will introduce three algorithms to solve these problems.

In Section 4.1, a new adaptive state estimation algorithm, namely adaptive fading Kalman filter (AFKF), is proposed to solve the divergence problem in state estimation using Kalman filter. It has been successfully applied to the headbox of a paper making machine for state estimation. In Section 4.2, an MIMO adaptive control strategy is developed for a paperboard machine to tackle the difficulties mentioned above. The control strategy incorporates conventional regulatory control technique, multivariable k-incremental predictor and self-tuning control algorithm. Section 4.3 investigates the application of model algorithmic control, which has the advantages of inherent time delay compensation, in paper machines. Feedforward decoupling is employed to eliminate the coupling in the process. Simulation studies on a real industrial paper machine have indicated the satisfactory performance of basis weight and moisture content control.

4.1 Adaptive Fading Kalman Filter

4.1.1 Introduction

Control system design very often involves the estimation of unmeasurable states. Kalman and Bucy (1961) introduced an effective algorithm to realize the optimum filter for Gaussian processes. The recursive computation nature

71

of the algorithm has attracted much attention. This well-known Kalman filtering technique has been widely employed in inertial navigation (Tze-Kwarfung and Grimble, 1983), target tracking (Chang and Tabaczynski, 1984) and industrial processes (Bialkowski, 1983).

In spite of its successful use, Kalman filter still remains some drawbacks. Inaccuracy in system models may seriously degrade the performance of the filter. Particularly, the usefulness of the filter may be nullified by the "divergence" phenomenon (Fitzgerald, 1971). The linear model of a real system is usually obtained as the results of either purposeful approximation and simplification or lack of knowledge about the true characteristics of the system, which is always erroneous. The convergence problem is hence becoming a main research subject of Kalman filter.

Shellenbarger (1966) considered the problem of unknown process noise covariances and proposed a maximum likelihood estimation of the unknown variable froms the residuals. Ohap and Stubberud (1976) provided an adaptive algorithm to determine the optimal gain matrix for discrete time systems with stationary ergodic white noise. Masreliez and Martin (1977), Tsai and Kurz (1983) compensated model errors by noises, and thus suggested the noise distributions be non-Gaussian. They proposed a robust Kalman filter based on the m-interval polynomial approximation (MIPA). Another approach to the divergence problem is to limit effective filter memory length. Fagin (1964), Sorenson and Sacks (1971) pointed out that a given linear model is adequate for a certain duration of time, but may be inadequate over long time intervals. Thus, they suggested to limit the memory of the Kalman filter by using exponential fading of past data via forgetting factors. On the other hand, it is advantageous to vary the time constant of the exponentially weighted filter when there are unpredictable jumps and drifts. Ydstie and Co (1985) proposed a variable forgetting factor (VFF) algorithm in which forgetting factors were determined based on "memory length". Rapid fading occurs when data fits the model poorly, and slow fading results in good fit.

Above techniques have successfully improved the convergence of Kalman filter (Xia, et al., 1994). However, there are still further needs to improve the optimality of filter. This section deals with the optimality and convergence of Kalman filters in the presence of both model parameters and noise covariance errors. New algorithms are developed to adaptively adjust the forgetting factors according to the optimality condition of the Kalman filter. Thus, the filter remains convergent and tends to be optimal in the cases where there exist model errors. The algorithms are efficient and have very moderate computation burden, and are thus convenient to be implemented for industrial applications.

4.1.2   Problem formulation

Consider a linear, discrete time, stochastic multivariable system

$$x(k+1) = \Phi(k+1, k)x(k) + G(k)w(k) \qquad (4.1)$$

$$y(k) = H(k)x(k) + v(k) \qquad (4.2)$$

where $x(k)$ is the $n \times 1$ state vector, $y(k)$ is the $m \times 1$ measurement vector, $\Phi(k+1, k)$ and $H(k)$ are state transition matrix and observation matrix, respectively. $w(k)$ and $v(k)$ denote sequences of uncorrelated Gaussian random vectors with zero means, the covariance matrices of which are

$$E[w(k)w^T(j)] = Q(k)\delta_{kj} \qquad (4.3)$$

$$E[v(k)v^T(j)] = R(k)\delta_{kj} \qquad (4.4)$$

The initial state $x(0)$ is specified as a random Gaussian vector

$$E[x(0)] = \bar{x}(0) \qquad E[(x(0) - \bar{x}(0))(x(0) - \bar{x}(0))^T] = P(0) \qquad (4.5)$$

If the system is completely observable, the equations describing the optimal estimator (the normal Kalman filter) are (Maybeck, 1982)

$$\hat{x}(k|k-1) = \Phi(k, k-1)\hat{x}(k-1) \qquad (4.6)$$

$$\hat{x}(k) = \hat{x}(k|k-1) + K(k)[y(k) - H(k)\hat{x}(k|k-1)] \qquad (4.7)$$

where

$$K(k) = P(k|k-1)H^T(k)[H(k)P(k|k-1)H^T(k) + R(k)]^{-1} \qquad (4.8)$$

$$P(k+1|k) = \Phi(k+1, k)P(k)\Phi^T(k+1, k) + G(k)Q(k)G^T(k) \qquad (4.9)$$

$$P(k) = [I - K(k)H(k)]P(k|k-1) \qquad (4.10)$$

The normal Kalman filter provides the best (minimum variance, unbiased) estimation $\hat{x}(k|k-1)$ of the state $x(k)$ with the given observations $y(k-1)$, $y(k-2)$, ..., $y(0)$ when the linear model for the system dynamics and measurement relation are perfect. Unfortunately, when the model is based on an erroneous model, the filter can "learn the wrong state too well" (Synder, 1973). Because the filter estimation highly depends upon the past data, and the system model degrades the measurement information from the distant past, the heavy reliance on the past data may cause state estimation to diverge. In order to overcome this problem, the filter should be capable of eliminating the effect of older data from a current state estimate if these data are no longer meaningful due to the erroneous model. Fagin (1964) initiated a method to limit the memory of the Kalman filter by using exponential fading of past data via forgetting factor $\lambda(k)$. The equations describing the fading Kalman filter are identical to those of the normal Kalman filter in equations (4.6)-(4.10) except the forgetting factor $\lambda(k)$ in the time propagation error covariance equation

$$P(k+1|k) = \lambda(k+1)\Phi(k+1,k)P(k)\Phi^T(k+1,k) + G(k)Q(k)G^T(k) \quad (4.11)$$

with $\lambda(k) \geq 1$. As a result, the influence of the most recent measured data in state estimation is overweighted and thus divergence is avoided.

The performance of the exponential fading Kalman filter fully depends on the selection of the forgetting factor. Therefore, how to generate optimal forgetting factor $\lambda(k)$ is the key problem in AFKF. In the following section, we present three algorithms for choosing optimal forgetting factor $\lambda(k)$ to improve the convergence and optimality of Kalman filter.

4.1.3   Main results

In developing the algorithms, we employ an important property of the optimal filter, that is, the residual $z(k)$ defined in the following equation is a white noise sequence when optimal filtering gain is used

$$z(k) = y(k) - H(k)\hat{x}(k|k-1) \qquad (4.12)$$

For an arbitrary gain $K(k)$, it can be shown that the covariance of the residual is

$$C_0(k) = E[z(k)z^T(k)] = H(k)P(k|k-1)H^T(k) + R(k) \qquad (4.13)$$

and the auto-covariance of the residual is

$$\begin{aligned}
C_j(k) &= E[z(k+j)z^T(k)] \\
&= H(k+j)\Phi(k+j,k+j-1)[I - K(k+j-1)H(k+j-1)]... \\
&\quad \times \Phi(k+2,k+1)[I - K(k+1)H(k+1)]\Phi(k+1,k) \\
&\quad \times [P(k|k-1)H^T(k) - K(k)C_0(k)] \quad \forall\, j = 1,2,3,... \qquad (4.14)
\end{aligned}$$

Substituting equations (4.8) and (4.13) into equation (4.14), $C_j(k)$ is identical to zero. This confirms that the sequence of residuals is uncorrelated when the optimal gain is used.

In practical situations, the real covariance of the residual $C_0(k)$ will be different from theoretical one given in equation (4.8)-(4.10) and (4.13) because of the errors in model parameters and noise covariances. Thus, $C_j(k)$ may not be identical to zero. From equation (4.14), we know that if a forgetting factor can be so chosen that the last term of $C_j(k)$, which is the only common term of $C_j(k)$ for all $j = 1, 2, \cdots$, be zero

$$P(k|k-1)H^T(k) - K(k)C_0(k) = 0 \qquad (4.15)$$

then $K(k)$ is optimal. In other words, if the gain is optimal, equation (4.15) holds. This forms the basis for the adaptive filtering algorithms developed below.

It should be noted that $C_0(k)$ in equation (4.15) is computed from measured data, rather than from equation (4.8)-(4.10) and (4.13).

Defining

$$S(k) = P(k|k-1)H^T(k) - K(k)C_0(k) \qquad (4.16)$$

the optimality of the Kalman filter can be judged by a scalar function defined as

$$f(\lambda; k) = \frac{1}{2} \sum_{i=1}^{n} \sum_{j=1}^{m} S_{ij}^2(k) \qquad (4.17)$$

where $S_{ij}(k)$ is the (i, j)th element of $S(k)$. The smaller the $f(k)$ is, the closer the filter is to the optimum. The absolute minimum of $f(k)$ means the most closely optimal estimation. Hence the forgetting factor $\lambda(k)$ should be chosen to minimize $f(k)$.

It should be pointed out that $C_j(k)$ depends also on the other terms in equation (4.14). However, these terms only include the gains in the future, $K(k+j-1)$ for $j = 2, 3, \cdots$. $C_1(k)$ depends just on the term $S(k)$. It is reasonable to consider equation (4.15) as a performance criterion, that is to adjust the current gain matrix to improve the performance of the filter.

Since the measurement matrix $H(k)$ is involved in optimality condition (4.15) and in the relation between $K(k)$ and $P(k|k-1)$ (equation (4.8)), we assume that $H(k)$ is perfect. This assumption is reasonable in most real world processes since measurement relations are usually much easier to obtain than the system dynamics.

Algorithm 1 (Steepest descent AFKF algorithm) *Given system equations (4.1) - (4.5), the optimal forgetting factor can be obtained through iterative computation of the equation*

$$\lambda^{l+1}(k) = \lambda^l(k) - \varphi \frac{\partial f^l(\lambda; k)}{\partial \lambda^l(k)} \qquad \forall l = 0, 1, 2, \ldots \qquad (4.18)$$

with initial conditions

$$\lambda^0(1) = 1 \qquad \lambda^0(k) = \lambda(k-1) \tag{4.19}$$

where k is the time series and l is the iteration times in a time instant.
$0 < \varphi < 1$ *is the step length in the gradient method.*

At the pth iteration, if the following condition holds

$$|\lambda^p(k) - \lambda^{p-1}(k)| < \varepsilon \tag{4.20}$$

stop iteration and take

$$\lambda(k) = max\{1, \ \lambda^p(k)\} \tag{4.21}$$

The gradient term in equation (4.18) is presented as

$$\frac{\partial f^l(\lambda; k)}{\partial \lambda^l(k)} = \sum_{i=1}^{n} \sum_{j=1}^{m} S_{ij}^l(k)(\frac{\partial S^l(k)}{\partial \lambda^l(k)})_{ij} \tag{4.22}$$

where

$$S^l(k) = P^l(k|k-1)H^T(k) - K^l(k)C_0(k) \tag{4.23}$$

$$
\begin{aligned}
\frac{\partial S^l(k)}{\partial \lambda^l(k)} &= \ \Phi(k, k-1)P(k-1)\Phi^T(k, k-1)H^T(k)\{I - [T^l(k)]^{-1}C_0(k)\} \\
&+ \ K^l(k)H(k)\Phi(k, k-1)P(k-1)\Phi^T(k, k-1)H^T(k) \\
&\quad \times \{I + [T^l(k)]^{-1}C_0(k)\}
\end{aligned}
\tag{4.24}
$$

and

$$P^l(k+1|k) = \lambda^l(k+1)\Phi(k+1, k)P(k)\Phi^T(k+1, k) + G(k)Q(k)G^T(k) \tag{4.25}$$

$$K^l(k) = P^l(k|k-1)H^T(k)[T^l(k)]^{-1} \tag{4.26}$$

$$T^l(k) = H(k)P^l(k|k-1)H^T(k) + R(k) \tag{4.27}$$

$$P^l(k) = [I - K^l(k)H(k)]P^l(k|k-1) \tag{4.28}$$

The value of $C_0(k)$ can be estimated using recursive equations

$$C_0(k) = G_1(k)/G_2(k) \tag{4.29}$$

$$G_1(k) = G_1(k-1)/\lambda(k-1) + z(k)z^T(k) \tag{4.30}$$

$$G_2(k) = G_2(k-1)/\lambda(k-1) + 1 \tag{4.31}$$

with initial conditions

$$G_1(0) = 0 \qquad G_2(0) = 0$$

Proof: The criterion $f(\lambda; k)$ given in equation (4.17) is a nonlinear function of $\lambda(k)$. The problem of searching optimal forgetting factor $\lambda(k)$ is equivalent to the problem of searching the absolute minimum of nonlinear function $f(k)$. Applying the steepest descent method, the forgetting factor can be calculated by

$$\lambda(k+1) = \lambda(k) - \varphi \frac{\partial f(\lambda; k)}{\partial \lambda(k)} \tag{4.32}$$

In order to find the gradient term in equation (4.32), taking derivations in equations (4.11) and (4.13) generates the following equations

$$\frac{\partial P(k|k-1)}{\partial \lambda(k)} = \Phi(k, k-1)P(k-1)\Phi^T(k, k-1) \tag{4.33}$$

$$\frac{\partial C_0(k)}{\partial \lambda(k)} = H(k)\frac{\partial P(k|k-1)}{\partial \lambda(k)}H^T(k) \tag{4.34}$$

Substituting equation (4.8) into (4.16) gives

$$S(k) = P(k|k-1)H^T(k)[I - T^{-1}(k)C_0(k)] \tag{4.35}$$

where $T(k)$ is defined in equation (4.27). Using equations (4.33) and (4.34) and differentiating equation (4.35) with respect to $\lambda(k)$, equation (4.24) (without the superscript l) is obtained.

From equation (4.17), equation (4.22) (without the superscript l) is obvious.

When more than one correction on $\lambda(k)$ are needed in a time instant, it is necessary to add superscript l in the related equations to indicate the number of iteration. Applying the above results and adding superscript l in equations (4.8), (4.10), (4.11), (4.34) and (4.35) forms equations (4.22)-(4.28).

To estimate $C_0(k)$ by using on-line measurement data, an unbiased consistent estimate for $C_0(k)$ based on k successive residuals is presented as

$$\bar{C}_0(k) = \frac{1}{k-1} \sum_{i=1}^{k} z(i)z^T(i) \tag{4.36}$$

Then a fading estimation formula for $C_0(k)$ can be given to overweight the most recent measured data

$$C_0(k) = \frac{\sum_{i=1}^{k-1} \sigma_{i,k} z(i)z^T(i) + z(k)z^T(k)}{\sum_{i=1}^{k-1} \sigma_{i,k} + 1} \tag{4.37}$$

where

$$\sigma_{i,k} = \prod_{j=i}^{k-1} \frac{1}{\lambda(j)} \tag{4.38}$$

Since $\lambda(j) \geq 1$ and thus $\sigma_{i,k} \geq \sigma_{i-1,k}$, it is obvious that the most recent data are overweighted. For the convenience of real-time application, we define

$$G_1(k) = \sum_{i=1}^{k-1} \sigma_{i,k} z(i) z^T(i) + z(k) z^T(k) \tag{4.39}$$

$$G_2(k) = \sum_{i=1}^{k-1} \sigma_{i,k} + 1 \tag{4.40}$$

It is easy to show that $G_1(k)$ and $G_2(k)$ can be computed recursively from equations (4.30) and (4.31). This completes the proof.

To improve the convergency and efficiency of the iteration process, Armijo algorithm (Polak, 1971) can be applied to choose step size.

Defining

$$C(\lambda^0; k) = \{\lambda | f(\lambda; k) \leq f(\lambda^0; k)\} \tag{4.41}$$

and

$$\theta(\varphi; \lambda; k) = [f(\lambda + \varphi h(\lambda); k) - f(\lambda; k)] - \langle \nabla f(\lambda; k), h(\lambda; k) \rangle \tag{4.42}$$

where $\nabla f(\lambda; k) = -\frac{\partial f(\lambda; k)}{\partial \lambda(k)}$, $h(\lambda; k)$ is given below. The Armijo algorithm can be stated as follows:

Step 1 Select $\lambda^0(k)$ such that the set $C(\lambda^0; k)$ is bounded; select $\alpha \in (0,1)$, $\beta \in (0,1)$ and $\rho > 0$.

Step 2 Set $l = 0$.

Step 3 Computer $h(\lambda^l; k) = -D(\lambda^l; k) \nabla f(\lambda^l; k)$.

Step 4 If $|h(\lambda^l; k)| \leq \epsilon$ (ϵ is a small positive scalar), set $\lambda(k) = \max\{1, \lambda^l(k)\}$ and stop; otherwise, goto next step.

Step 5 Set $\mu = \rho$.

Step 6 Compute $\theta(\mu; \lambda^l; k)$.

Step 7 If $\theta(\mu; \lambda^l; k) \leq 0$, set $\varphi^l = \mu$ and goto next step; otherwise set $\mu = \beta \mu$ and goto step 6.

Step 8 Set $\lambda^{l+1}(k) = \lambda^l(k) - \varphi^l h(\lambda^l; k)$; set $l = l + 1$ and goto step 3.

In the algorithm, we select $D(\lambda^l; k) = 1$, $\lambda^0(1) = 1$, $\lambda^0(k) = \lambda(k-1)$, $\alpha = \frac{1}{2}$, $\beta \in (0.5, 0.8)$ and $\rho = 1$.

Algorithm 1 fails to give an explicit formula for the calculation of $\lambda(k)$. Since iterative computation is involved, it may be difficult to apply this algorithm to real-time processes. In the remainder of this section, two one-step algorithms are developed.

Algorithm 2 (One-step AFKF algorithm) *Given system state equations (4.1)-(4.5) with the following assumptions (I) and (II):*
(I) $Q(k)$, $R(k)$ and $P(0)$ are all positive definite;
(II) The measurement matrix $H(k)$ is full-ranked.
The optimal forgetting factor can be computed by

$$\lambda(k) = max\{1, \frac{1}{m}trace[N(k)M^{-1}(k)]\} \tag{4.43}$$

where

$$M(k) = H(k)\Phi(k, k-1)P(k-1)\Phi^T(k, k-1)H^T(k) \tag{4.44}$$

$$N(k) = C_0(k) - H(k)G(k-1)Q(k-1)G^T(k-1)H^T(k) - R(k) \tag{4.45}$$

Proof: Substituting equation (4.8) into the optimality condition (4.15) gives

$$P(k|k-1)H^T(k)\{I - [H(k)P(k|k-1)H^T(k) + R(k)]^{-1}C_0(k)\} = 0 \tag{4.46}$$

Since $P(k|k-1)$ is nonsingular and $H(k)$ is assumed to be full-ranked, it is obvious that equation (4.46) implies the following relationship

$$[H(k)P(k|k-1)H^T(k) + R(k)]^{-1}C_0(k) = I \tag{4.47}$$

or

$$H(k)P(k|k-1)H^T(k) = C_0(k) - R(k) \tag{4.48}$$

Equation (4.48) means that with assumptions (I) and (II), the optimality condition (4.15) is equivalent to (4.13). Substituting equation (4.11) into (4.48) and reorganizing it generates

$$\lambda(k)H(k)\Phi(k, k-1)P(k-1)\Phi^T(k, k-1)H^T(k)$$
$$= C_0(k) - H(k)G(k-1)Q(k-1)G^T(k-1)H^T(k) - R(k) \tag{4.49}$$

Using equations (4.44) and (4.45), equation (4.49) is simplified as

$$\lambda(k)M(k) = N(k) \tag{4.50}$$

or

$$\lambda(k)I = N(k)M^{-1}(k) \qquad \lambda(k) \geq 1 \tag{4.51}$$

Taking trace in both sides of the equation, we get (4.43), and this completes the proof.

In equation (4.43), the inversion of matrix $M(k)$ is involved, which will complicate the computation. To avoid inversion manipulation, trace is directly taken in both sides of equation (4.50) and this gives Algorithm 3.

Algorithm 3 (Simplified one-step AFKF algorithm) *Using system state equations and conditions given in Algorithm 2, the optimal forgetting factor can be computed by the following equation*

$$\lambda(k) = max\{1,\ trace[N(k)]/trace[M(k)]\} \tag{4.52}$$

where matrices $M(k)$ and $N(k)$ are defined by equations (4.44) and (4.45).

In choosing the preferable algorithm, the tradeoff between performance and real-time computational effort should be considered. In fact, simulation and industrial application results show that the performance of Algorithm 3 is quite satisfactory.

From equations (4.43)-(4.45) and (4.52), the physical meanings of optimal forgetting factor is clear: for unknown drifts and process changes, the adaptive fading algorithm compensates the increasing estimation errors by choosing larger forgetting factor.

The uniformly asymptotical stability of AFKF, as given in the following theorem, can be obtained by applying the results obtained by Deyst and Price (1968), as well as Sorenson and Sacks (1971).

Theorem 1 *Assume system described by (4.1) and (4.2) is stochastically controllable and observable, which implies that there are real numbers α_1, α_2, β_1 and β_2 such that the conditions (Deyst and Price, 1968)*

$$\alpha_1 I \leq \sum_{i=k-N}^{k-1} \Phi(k,i+1)G(i)Q(i)G^T(i)\Phi^T(k,i+1) \leq \alpha_2 I \tag{4.53}$$

$$\beta_1 I \leq \sum_{i=k-N}^{k} \Phi(i,k)H^T(i)R(i)^{-1}H(i)\Phi(i,k) \leq \beta_2 I \tag{4.54}$$

holds for all $k \geq N$, then the adaptive fading Kalman filter is uniformly asymptotically stable in the large.

Proof: Using equation (4.10), the normal Kalman filter described in equations (4.6) and (4.7) can be rewritten as

$$\hat{x}_n(k) = F_n(k,k-1)\hat{x}_n(k-1) + D_n(k)y(k) \tag{4.55}$$

where

$$F_n(k,k-1) = P_n(k)P_n^{-1}(k|k-1)\Phi(k,k-1) \tag{4.56}$$

$$D_n(k) = P_n(k)H^T(k)R^{-1}(k) \tag{4.57}$$

Applying Deyst and Price's (1968) results, it follows that the matrix $F_n(k, k-1)$ describes a system which is asymptotically stable in large. Since equation (4.11) can be rewritten into the following form

$$P(k+1|k) = \bar{\Phi}(k+1, k)P(k)\bar{\Phi}^T(k+1, k) + G(k)Q(k)G^T(k) \tag{4.58}$$

where

$$\bar{\Phi}(k+1, k) = [\lambda(k+1)]^{\frac{1}{2}}\Phi(k+1, k) \tag{4.59}$$

This implies that $P(k)$ and $P(k|k-1)$ can be considered as the estimation error covariance matrices of the normal Kalman filter for the modified system

$$x(k+1) = \bar{\Phi}(k+1, k)x(k) + G(k)w(k) \tag{4.60}$$

Provided $1 \le \lambda(k+1) < \infty$, the system (4.60) and (4.2) is stochastically controllable and observable if and only if the system (4.1) and (4.2) is stochastically controllable and observable. Therefore, the matrix

$$\bar{F}_n(k, k-1) = P(k)P^{-1}(k|k-1)\bar{\Phi}(k, k-1) \tag{4.61}$$

is uniformly asymptotically stable.

Similarly, the adaptive fading Kalman filter described in equations (4.6)-(4.8), (4.10) and (4.11) can be rewritten as

$$\hat{x}(k) = F(k, k-1)\hat{x}(k-1) + D(k)y(k) \tag{4.62}$$

where

$$F(k, k-1) = P(k)P^{-1}(k|k-1)\Phi(k, k-1) \tag{4.63}$$

$$D(k) = P(k)H^T(k)R^{-1}(k) \tag{4.64}$$

It is found from equations (4.61) and (4.63) that

$$F(k, k-1) = \bar{F}_n(k, k-1)/[\lambda(k)]^{\frac{1}{2}} \tag{4.65}$$

Since $1 \le \lambda(k) < \infty$ for all k and $\bar{F}_n(k, k-1)$ is uniformly asymptotically stable, the optimal fading Kalman filter which is described by $F(k, k-1)$ must be also uniformly asymptotically stable and the proof is completed.

Remarks:

(1) When the filtering model is exactly correct, $C_0(k)$ is given as equation (4.13). Substituting equation (4.13) into (4.44) and (4.45), we have $M(k) =$

$N(k)$, and thus results in $\lambda(k) = 1$. This implies that in the case where exact filtering model is used, AFKF functions the same as the normal Kalman filter and provides the optimal (minimum variance, unbiased) estimation. When the model errors cause the covariance of the residual to deviate from the theoretical one, AFKF compensates the model errors by adjusting the forgetting factor according to the optimality condition. In this way, the filter achieves better performance.

(2) The matrix $S(k)$ has $n \times m$ elements. Since only one factor $\lambda(k)$ can be used to null out terms in $S(k)$, the filter may not be exactly optimal. However, since $\lambda(k)$ is adjusted according to the optimality condition of the filter, the optimality and convergence of the Kalman filter are surely improved.

4.1.4   Simulation studies

The comparison of performance between normal Kalman filter and AFKF has been undertaken in the cases of model coefficient errors and unknown drifts.

For simplicity, one-dimensional random state model is considered. For the scalar system, Algorithms 2 and 3 give the same results.

Case 1. Model coefficient errors

The system equations of scalar discrete-time random state $x(k)$ are represented as

$$x(k + 1) = 0.4x(k) + w(k) \tag{4.66}$$

$$y(k) = x(k) + v(k) \tag{4.67}$$

where $w(k) \sim N(0, 0.1^2)$, $v(k) \sim N(0, 0.5^2)$ and $x(0) \sim N(2, 0.2^2)$, and the erroneous filtering model is

$$x(k + 1) = 0.8x(k) + w(k) \tag{4.68}$$

$$y(k) = x(k) + v(k) \tag{4.69}$$

The simulation results are presented in Figure 4.1.

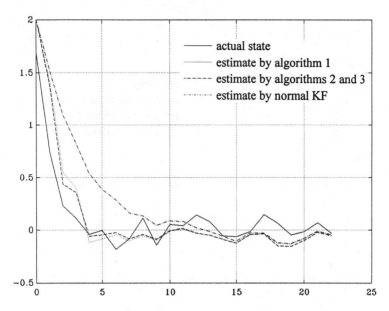

Figure 4.1: Simulation results with system model errors

Case 2. Unknown drifts

Consider a random state described by equations

$$x(k+1) = 0.5x(k) + 0.4 + w(k) \tag{4.70}$$

$$y(k) = x(k) + v(k) \tag{4.71}$$

where $w(k) \sim N(0, 0.05^2)$, $v(k) \sim N(0, 0.05^2)$ and $x(0) \sim N(2, 0.05^2)$, and the erroneous filtering model is

$$x(k+1) = 0.5x(k) + w(k) \tag{4.72}$$

$$y(k) = x(k) + v(k) \tag{4.73}$$

The simulation results are presented in Figure 4.2.

From Figures 4.1 and 4.2, it is found that the simulation results agree with the theoretical ones. During the initial period, a larger forgetting factor is generated due to the poor fit of the model with the actual process. Normal Kalman filter starts to degrade, but AFKF still performs well through adaptively adjusting the forgetting factors. After a while, the system closes to its new steady state, and the difference between the actual state and the estimate becomes smaller. As a result, the forgetting factor returns to its normal value,

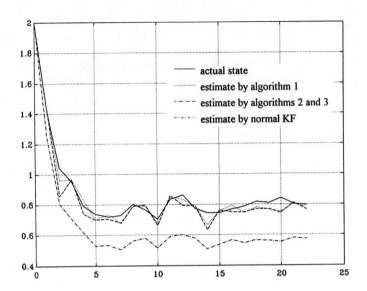

Figure 4.2: Simulation results with unknown drifts

1. Figure 4.3 shows the time evolution of the forgetting factors for Case 1. Sorenson and Sacks (1971) indicated that if the model discrepancy is very significant, the fading factor will be very large in order to recover the filter from divergence. The forgetting factors for Case 2 have the similar behavior. As predicted theoretically, AFKF has satisfactory dynamic performance, and the residuals of state estimate are almost eliminated. It can also be found that the performance of Algorithms 1, 2 and 3 are nearly the same.

4.1.5   Application to paper machine headbox

The AFKF algorithm has been applied to a real industry paper machine, which produces super-thin condenser paper. The headbox section, as shown in Figure 4.4, is an essential part of the paper machine. Since this paper machine has slow machine rate (70 m/min), its headbox is not pressurized, but open to atmosphere. The main purpose of the headbox is to distribute the water/fiber suspension onto the wire as evenly as possible. Thick stock from the machine chest is diluted by white water to form thin stock which then flow onto the wire. The flow rate of both thick stock and white water are controlled.

The dry basis weight of paper sheet on reel is an important quality variable, which varies with the flow rate and the consistency of stock onto the wire.

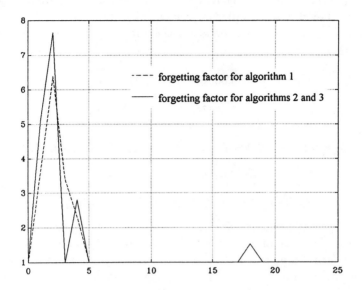

Figure 4.3: Time evolution of the forgetting factor in Case 1

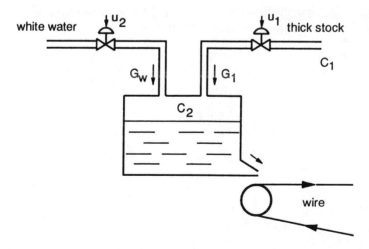

Figure 4.4: Principle diagram of headbox section

However, the measurement sensor for basis weight is not available on the paper machine. The purpose of using AFKF is to obtain an estimate of dry basis weight and other states to implement control algorithm.

The model of the headbox section of the paper machine can be described by (Xia, 1989)

$$G_1(k+1) = 0.8667G_1(k) - 0.0344u_1(k-1) \tag{4.74}$$

$$G_w(k+1) = 0.8667G_w(k) - 0.6877u_2(k-1) \tag{4.75}$$

$$\begin{aligned} C_2(k+1) &= 0.9099C_2(k) + 0.1069G_1(k) - 0.0503G_w(k) \\ &+ 0.0788C_1(k-2) \end{aligned} \tag{4.76}$$

$$B_w(k) = 4.089G_1(k-3) + 1.4910G_w(k-3) + 1.7533C_2(k-3) \tag{4.77}$$

where C_1, C_2, G_1, G_w and B_w are the consistency of thick stock, consistency of thin stock, flow rate of thick stock, flow rate of white water, and dry basis weight, respectively. u_1 and u_2 are the changes of the openings of thick stock and white water control valves. The sampling interval is 20 seconds, and G_1, C_1 and C_2 can be measured on-line.

Denoting $x(k) = (G_1(k)\ G_w(k)\ C_2(k))^T$, $y(k) = (G_1(k)\ C_2(k))^T$, $u(k) = (u_1(k-1)\ u_2(k-1))^T$ and $d(k) = C_1(k-2)$ as state vector, output vector, control vector and disturbance vector of the system, respectively, and adding input noise $w(k)$ and measurement noise $v(k)$ to the model, then the state space model of headbox section can be written as

$$\begin{aligned} x(k+1) &= \begin{pmatrix} 0.8667 & 0 & 0 \\ 0 & 0.8667 & 0 \\ 0.1069 & -0.0503 & 0.9099 \end{pmatrix} x(k) \\ &+ \begin{pmatrix} -0.0344 & 0 \\ 0 & -0.6877 \\ 0 & 0 \end{pmatrix} u(k) \\ &+ \begin{pmatrix} 0 \\ 0 \\ 0.0788 \end{pmatrix} d(k) + w(k) \end{aligned} \tag{4.78}$$

$$y(k) = \begin{pmatrix} 1 & 0 & 0 \\ 0 & 0 & 1 \end{pmatrix} x(k) + v(k) \tag{4.79}$$

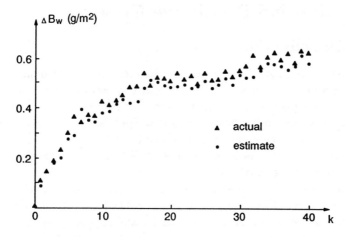

Figure 4.5: Basis weight with step change in thick stock valve

Algorithm 3 is used to estimate the dry basis weight of paper sheet on reel. Figure 4.5 shows the change of actual and estimated basis weights when the opening of thick stock valve is increased steply. Since there is no on-line measurement for basis weight, the actual basis weight is obtained by manually testing the paper sheet samples picked up from the dry end of the paper machine. The filter chooses covariances $Q(k) = diag(0.4^2, 0.4^2, 1)$, $R(k) = diag(0.3^2, 0.4^2)$, $P(0) = diag(0.8^2, 0.8^2, 0.8^2)$. It is seen that the actual and estimated basis weights match very well. AFKF has also been applied to a basis weight and moisture control system for the paper machine. The computer control system has been in operation reliably and satisfactorily.

4.2   Adaptive Predictive Control

4.2.1   Introduction

Basis weight and moisture content are the most important quality variables of paper products. A good control of them can significantly improve paper quality, increase production rate, and reduce raw material and energy consumption. Previous studies have shown that applications of advanced process control are of good economic returns (Shead, 1979; Wallace, 1980). Typical returns have been in the order of decreasing fiber 5% and steam consumption 10%, and increasing annual production 5%-10%.

The basis weight and moisture content in a paperboard machine are difficult to control due to the following reasons: (1) Paperboard machines have great parameter drifts, external disturbances and random noises. Some of the external disturbances are measurable and others are not; (2) The dynamics of the paper machine are nonlinear and time varying; (3) There exist large time delays (dead times) in basis weight and moisture content control loops; (4) There are serious interactions between the two control loops. The first two problems cause time and operating condition dependent changes in the process dynamics. The last two imply that a multivariable controller is required to deal with multiple large time delays.

Adaptive control has long been considered as an ideal solution to the control problem of time varying stochastic systems. Its ability to adapt itself to changing process behavior, particularly the changes caused by the underlying nonlinearity of process dynamics, makes the adaptive control approach very appealing. Åström initiated the application of self-tuning regulator to paper machines (Åström, 1967). Since then, a number of adaptive control techniques have been applied to pulp and paper processes (Allison, et al., 1991; Åström, et al., 1977; Cegrell and Hedquist, 1975; Sikora and Bialkowski, 1984; Wilhelm, 1982).

This section discusses the development of a general multi-input/multi-output (MIMO) adaptive controller for paperboard machines with multiple time delays, measurable and unmeasurable disturbances. The model of the paperboard machine is identified by a recursive forgetting algorithm. A k-incremental predictor is applied to eliminate the steady state errors in parameter identification and control. As a result, the robustness of the control system against machine rate changes and unmeasurable disturbances as well as stochastic noises are greatly improved. An auxiliary control variable is introduced to deal with the time-delay difference between the two control loops.

4.2.2   Fundamentals of the paperboard machine

The paperboard machine investigated is a cylinder-mould machine in a paper mill, which produces packing board using straw pulp and wood pulp. Figure 4.6 presents the flow diagram of the paperboard machine.

The straw pulp and wood pulp are pumped into high-level tanks and diluted with water, from where they flow through sand tray and rotary screens to clean impurities. The diluted and cleaned pulp are sent to the distributor and then distributed into four straw pulp flows and one wood pulp flow. These five pulp flows to the five wire-coverd cylinders (forming cylinders). The paper web is formed on the surface of wire-covered cylinders in the cylinder forming vats. The vats are partially filled with diluted fiber suspensions. As the forming cylinder rotates through the vat of stock, the water flows through the surface of the cylinder and the fibers catch on the wire, forming the web of paper.

A moving felt is pressed against the wet web and removes the wet web from the forming cylinder at this point. The felt and web picked up from the first cylinder proceed to the second cylinder, where they are pressed against the web formed there to pick it up. The felt continues through the forming section, picking up webs from all the five vats in succession. After all webs are picked up by the felt and are combined to form a thick composite board web, it goes through the press section. The web which leaves the press section end is passed around a series of steam-filled cylinders where the remaining water is removed by evaporation. The web travels in such a way that it passes over the top and under the bottom cylinders, so that first one side of the web, then the other, is heated.

Since the wood pulp forms a web on the top surface of the paperboard, it is called surface stock, whereas the straw pulp is called inner stock.

The operation of the paperboard machine requires the basis weight and moisture content to meet process specifications while minimizing the energy and pulp consumption per ton of paperboard product. The basis weight is controlled by the straw pulp control valve V_s, while the moisture content is controlled by the steam control valve V_p.

The paperboard machine used to be controlled by two single-loop PID regulators which maintained the pulp flow rate and steam pressure constant. Operation experience showed the reel basis weight and moisture content presented large variations under the control of single-loop regulators.

The difficulties in the paperboard machine control arise from:

(1) The machine produces packing paperboard of basis weight as high as $360g/m^2$. However, the machine speed is as low as $50m/min$. Greater thickness of the web requires more drying cylinders. There are forty-five drying

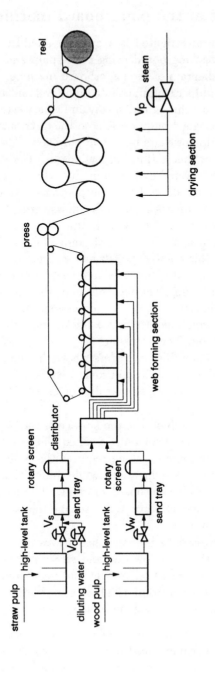

Figure 4.6: Flow diagram of the paperboard machine

cylinders in this paperboard machine. It results in a large time delay. The time delay from the straw pulp control valve to reel basis weight is about 6 minutes. The time delay form the steam control valve to reel moisture content is about 3 minutes.

(2) There are serious couplings between the control of basis weight and moisture content. If moisture content increases by 1%, the basis weight will increase by 4g/m^2. Similarly, change in basis weight has great influence on moisture content. The couplings make the paperboard machine be a multi-variable system. The system could be triangularized by considering moisture content and dry basis weight. However, since the paper company considered the basis weight rather than the dry basis weight as a quality variable, the operators required the system to control the basis weight directly.

(3) The capability of the forming cylinders to catch fibers varies with the wear of wires and felts. The variation of machine speed also changes the dynamics of the paperboard machine.

Considering the above problems, we propose the following multivariable adaptive controller for basis weight and moisture content control.

4.2.3 Control strategy

The basis weight and moisture content are affected by many variables. Some of them are measurable, such as the consistencies of straw pulp and wood pulp, machine speed, steam pressure, etc. Whereas others are unmeasurable, such as the quality of pulps, the capability of forming cylinders to catch fibers, the stock level of vats, and the efficiency of the press and coating, etc. Due to the large time delays, some auxiliary feedback and feedforward control loops are required to maintain the basis weight and moisture content within the specifications.

The structure of the basis weight and moisture content control strategy is shown in Figure 4.7. The important features of this strategy are:

(1) The consistency of the straw pulp from the pulp preparation section varies widely and has a large influence on the reel basis weight. To kept it constant, the consistency is controlled by a PID controller through adjusting the flow rate of the diluting water.

(2) Another PID controller is used to control the steam pressure.

(3) The measurements of the straw pulp and wood pulp consistency serve as feedforward signals to adjust the opening of the control valves V_s and V_p,

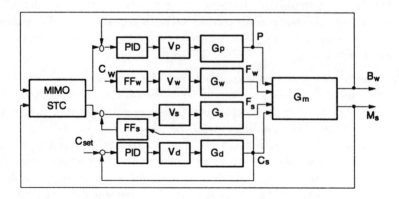

Figure 4.7: Adaptive control strategy for the paperboard machine

respectively. In this way, the variation of pulp consistencies is compensated before it finally affects the paperboard quality.

(4) The multivariable self-tuning controller provides the setpoint for the PID steam pressure controllers and the opening of the straw pulp control valve V_s.

The control strategy has the following advantages: despite the large time delay in basis weight and moisture control loops, the effects of the variation of straw pulp and wood pulp consistencies, and the steam pressure on final paperboard can be eliminated quickly. The self-tuning controller is used to compensate unmeasurable disturbances.

4.2.4   Self-tuning controller

Self-tuning controllers combine a recursive parameter estimator and a control design mechanism. The parameter estimator obtains a plant model from input/output data. The control design mechanism accepts the parameter estimates as if they were exact and from which the parameters of a feedback control law can be deduced. A self-tuning control system provides the capability of automatically compensating for parameter and environmental variations that may occur during operation. In the case of a paperboard machine, such a system has the potential for providing uniform basis weight and moisture content despite the process dynamics variation and external disturbances.

Since the middle of 1970s, many researchers have been investigating multivariable self-tuning controllers. Borisson (1975) proposed a self-tuning controller for MIMO system. The controller was only applied to systems with equal numbers of inputs and outputs, or square systems. Koivo and Tanttu (1975) derived a self-tuning controller for the case where the number of outputs is not necessarily equal to the number of inputs, i.e., non-square systems.

Many simulation studies have concentrated on the dynamics and data which have zero mean so that no offsets are obtained. However, the input/output data supplied to the parameter estimator are very often insufficient to provide an accurate value of the effective additive constant in the plant model. To deal with non-zero mean disturbance and offset problems, Clarke and his co-workers proposed a novel k-incremental predictor, in which the control law and identification use the differences of inputs and outputs instead of inputs and outputs themselves. Digital simulation and application studies have shown the advantages of the k-incremental predictor (Clarke, et al., 1983).

In the basis weight and moisture content control system of the paperboard machine, we mainly apply the multivariable self-tuning controller developed by Koivo and Tanttu (1985), and make a number of modifications:

(1) The k-incremental predictor (Clarke, et al., 1983) is generalized into multivariable case and applied to self-tuning controller. The purpose is to eliminate the offset terms in parameter estimation and control algorithm, thus to improve the regulation against non-zero disturbances such as those induced by load changes.

(2) An auxiliary control variable is introduced to solve the problem arising from the difference of time delay in two control loops.

Multivariable k-incremental predictor and self-tuning controller

The linearized model of the controlled process is assumed to be

$$A(z^{-1})y(t) = B(z^{-1})z^{-k}u(t) + C(z^{-1})e(t) + d \qquad (4.80)$$

where $k \geq 1$ is time-delay of the system, $y(t)$ and $u(t)$ are r-dimensional output vector and input vector, respectively. The operator matrices $A(z^{-1})$, $B(z^{-1})$ and $C(z^{-1})$ are defined as follows

$$A(z^{-1}) = I + A_1 z^{-1} + A_2 z^{-2} + \cdots + A_n z^{-n}$$

$$B(z^{-1}) = B_0 + B_1 z^{-1} + B_2 z^{-2} + \cdots + B_n z^{-n}, \quad B_0 \neq 0$$

$$C(z^{-1}) = I + C_1 z^{-1} + C_2 z^{-2} + \cdots + C_n z^{-n}$$

where all zeros of $B(z^{-1})$ and $C(z^{-1})$ lie strictly outside the unit disc, that is, the solutions of $B(z^{-1}) = 0$ and $C(z^{-1}) = 0$ satisfy $|z| < 1$. $e(t)$ is a sequence of independent $(0, \Omega)$ variables representing the stochastic part of the disturbance. d is an unknown constant vector which arises from:

(i) load disturbances, such as changes in the consistency and flow rate of straw pulp and wood pulp;

(ii) the fact that the steady-state incremental gain $\partial y / \partial u$ of a process does not equal to the static gain y/u – a consequence of the large signal nonlinearity of most plant;

(iii) non-zero mean noise.

Our task is to find a control vector $u^*(t)$ that minimizes the cost function

$$I_p = E\{\| \ P(z^{-1})y(t+k) - R(z^{-1})w(t) + Q'(z^{-1})u(t) \ \|^2\} \tag{4.81}$$

where $w(t) \in R^r$ is the reference signal sequence, $P(z^{-1})$, $R(z^{-1})$ and $Q'(z^{-1})$ are $(r \times r)$-dimensional polynomial matrices with

$$P(z^{-1}) = P_0 + P_1 z^{-1} + \cdots + P_p z^{-p}$$

$$R(z^{-1}) = R_0 + R_1 z^{-1} + \cdots + R_l z^{-l}$$

$$Q'(z^{-1}) = Q'_0 + Q'_1 z^{-1} + \cdots + Q'_s z^{-s}$$

We define an auxiliary output $\Phi(t)$ by

$$\Phi(t+k) := Py(t+k) - Rw(t) + Qu(t) \tag{4.82}$$

where $Q = ((P_0 B)^{-1}(Q'_0))^T Q'$. To derive an optimal predictor, consider the identity

$$PC = AE + z^{-k}F \tag{4.83}$$

Given A, P, C and k, the coefficients of $E(z^{-1})$ and $F(z^{-1})$ polynomials

$$E(z^{-1}) = E_0 + E_1 z^{-1} + \cdots + E_{k-1} z^{-k+1}$$

$$F(z^{-1}) = F_0 + F_1 z^{-1} + \cdots + F_{n_F} z^{-n_F}$$

where

$$n_F = max(n-1, \ p+n-k)$$

can be uniquely determined from equation (4.83). Denoting $\tilde{E}$ and $\tilde{F}$ as the left and right coprime factorization of E and F, that is

$$\tilde{E}F = \tilde{F}E \tag{4.84}$$

with

$$det\ \tilde{E} = det\ E, \qquad det\ \tilde{F} = det\ F$$

and defining a polynomial matrix $\tilde{C}(z^{-1})$ such that

$$\tilde{C}P = \tilde{E}A + z^{-k}\tilde{F} \tag{4.85}$$

from equations (4.83) - (4.85), we obtain

$$\tilde{C}E = \tilde{E}C \tag{4.86}$$

Premultiplying equation (4.80) by $\tilde{E}$ and using equations (4.85) and (4.86) gives

$$\tilde{C}Py(t+k) = \tilde{F}y(t) + \tilde{E}Bu(t) + \tilde{E}d + \tilde{C}Ee(t+k) \tag{4.87}$$

Hence, the prediction model is

$$\tilde{C}[Py(t+k|t)]^* = \tilde{F}y(t) + \tilde{E}Bu(t) + \tilde{E}d \tag{4.88}$$

where $[Py(t+k|t)]^*$ is the prediction of $Py(t+k|t)$ given data up to t. The prediction error is given by

$$\zeta(t+k) = y(t+k) - [Py(t+k|t)]^* = Ee(t+k) \tag{4.89}$$

Without the loss of generality, we assume that the noise matrix C equals to identity. It can be shown (Koivo and Tanttu, 1985) that $\tilde{C} = I$. Defining $\tilde{E}d = d_1$, equation (4.88) can be rewritten as

$$[Py(t+k|t)]^* = \tilde{F}y(t) + \tilde{E}Bu(t) + d_1 \tag{4.90}$$

or equivalently

$$[Py(t+k|t)]^* = [Py(t|t-k)]^* + \tilde{F}\Delta_k y(t) + \tilde{E}B\Delta_k u(t) \tag{4.91}$$

where

$$\Delta_k y(t) = y(t) - y(t-k) \qquad \Delta_k u(t) = u(t) - u(t-k) \tag{4.92}$$

From equation (4.89), we have

$$[Py(t|t-k)]^* = Py(t) - \zeta(t) \tag{4.93}$$

Substituting the above equation into (4.91) gives the multivariable k-incremental predictor

$$[Py(t+k|t)]^* = Py(t) - \zeta(t) + \tilde{F}\Delta_k y(t) + \tilde{E}B\Delta_k u(t) \tag{4.94}$$

The optimal k-incremental prediction of the auxiliary output $\Phi(t + k)$ is then given as:

$$\Phi^*(t + k) = Py(t) - \zeta(t) + \tilde{F}\Delta_k y(t) + \tilde{E}B\Delta_k u(t) - Rw(t) + Qu(t) \quad (4.95)$$

It can be shown that the optimal control $u^*(t)$ which minimizes the cost function I_p can be solved from the identity (Koivo and Tanttu, 1985)

$$\Phi^*(t + k) = 0 \quad (4.96)$$

the self-tuning control law is given

$$(Q + \tilde{G}\Delta_k)u(t) = Rw(t) - (P + \tilde{F}\Delta_k)y(t) + \zeta(t) \quad (4.97)$$

where $\tilde{G} := \tilde{E}B$.

Parameter estimation

Let us briefly discuss the parameter updating for easier reference. From equations (4.89) and (4.94), we have

$$Py(t) = Py(t - k) - \zeta(t - k) + \tilde{F}\Delta_k y(t - k) + \tilde{G}\Delta_k u(t - k) + \zeta(t) \quad (4.98)$$

Define

$$\Psi := [\psi_1 \psi_2 \cdots \psi_r]^T = [P_0\ P_1\ \cdots\ P_p]$$
$$\zeta(t) = [\zeta_1(t)\ \zeta_2(t)\ \cdots\ \zeta_r(t)]^T$$

$$X(t - k) := \{[y(t - k) - y(t - 2k)]^T\ [y(t - k - 1) - y(t - 2k - 1)]^T$$
$$\cdots\ [y(t - k - m) - y(t - 2k - m)]^T$$
$$[u(t - k) - u(t - 2k)]^T\ [u(t - k - 1) - u(t - 2k - 1)]^T$$
$$\cdots\ [u(t - k - q) - u(t - 2k - q)]^T\}$$

$$Y(t) := \{[y(t) - y(t - k)]^T\ [y(t - 1) - y(t - k - 1)]^T\ \cdots\ [y(t - p) - y(t - k - p)]^T\}$$
$$\Theta := [\theta_1\ \cdots\ \theta_r]^T = [\tilde{F}_0\ \tilde{F}_1 \cdots \tilde{F}_m\ \tilde{G}_0\ \tilde{G}_1 \cdots \tilde{G}_q]$$
$$Z_i(t) := \psi_i Y(t)$$

where p, m and q are the order of $P(z^{-1})$, $\tilde{F}(z^{-1})$ and $\tilde{G}(z^{-1})$, respectively. Equation (4.98) can be rewritten as

$$Z_i(t) = X(t - k)\theta_i + \zeta_i(t - k) + \zeta_i(t), \quad i = 1, 2, \cdots, r \quad (4.99)$$

The controller parameters can be estimated with the standard recursive least square algorithm

$$\hat{\theta}_i(t) = \hat{\theta}_i(t-1) + K_i(t)[Z_i(t) - X(t-k)\hat{\theta}_i(t-1)] \qquad (4.100)$$

$$K_i(t) = [\lambda_i + X(t-k)V_i(t-1)X^T(t-k)]^{-1}V_i(t-1)X^T(t-k) \qquad (4.101)$$

$$V_i(t) = [V_i(t-1) - K_i(t)X^T(t-k)V_i(t-1)]/\lambda_i \qquad (4.102)$$

where λ_i is the forgetting factor.

The self-tuning control algorithm is obtained directly as follows:

Step 1 Read the new set point $w(t)$ and output $y(t)$.

Step 2 Compute the offset $\zeta_i(t)$

$$\hat{\zeta}_i(t) = Z_i(t) - X(t-k)\hat{\theta}_i(t-1) - \hat{\zeta}_i(t-k)$$

Step 3 Estimate the controller parameters using equations (4.100)-(4.102).

Step 4 Compute the new control from equation (4.97).

Step 5 Set t:=t+1 and return to step 1.

4.2.5 Implementation issues

Implementing the self-tuning controller on the paperboard machine raises a number of practical issues. Among them is the identification of system dynamics. Although the self-tuning controller is not sensitive to model parameter errors, the model structure or model order has an important impact on system performance. Besides, a relatively accurate initial model can avoid the improper control action. Higher order models can offer the advantage of higher modelling accuracy, however, increase the computation load of the parameter estimation algorithm. Therefore, a compromise between model accuracy and computation burden is required.

The recursive parameter estimation algorithm provided for scalar controlled ARMA processes (Mayne, et al., 1984) is extended to multivariable systems and applied to obtain the initial model of the paperboard machine. The input signal for identification is a pseudo-random binary sequence (PRBS). Since paperboard machine is a 2-input/2-output system, the two input signals should be linearly independent in order to obtain an accurate model. We generate a PRBS of length N_p. The first $(N_p - 1)/2$ numbers serve as first input, and the second $(N_p - 1)/2$ numbers serve as second input. Based on the knowledge about the process dynamics, we choose the PRBS of length $N_p = 127$, interval $\Delta = 2.0min$, amplitude $a_1 = 1.5mA$ and $a_2 = 0.3mA$, and sampling time $T_s = 1.0min$.

The models obtained from system identification are

$$(1.0 - 1.15z^{-1} + 0.33z^{-2})y_1(t) = (6.1 - 3.36z^{-1})z^{-6}u_1(t)$$
$$-(3.4 - 2.03z^{-1})z^{-3}u_2(t) + e_1(t)$$

$$(1.0 - 1.32z^{-1} + 0.43z^{-2})y_2(t) = (0.4 - 0.25z^{-1})z^{-6}u_1(t)$$
$$-(1.20 - 0.91z^{-1})z^{-3}u_2(t) + e_2(t)$$

where y_1 and y_2 are basis weight and moisture content, respectively. u_1 and u_2 are positions of straw pulp control valve and the value of steam pressure.

Since the basis weight and moisture content control loops have different time delays, normal multivariable self-tuning control algorithms fail to provide a feasible solution. To solve the problem, an auxiliary control variable $u_2'(t)$ is introduced

$$u_2'(t) := u_2(t - 3)$$

which means that the control action for steam control valve will be delayed for three sampling intervals before it is issued. This artificially introduced delay may deteriorate the performance of moisture content control. In the practical implementation it has been found that the two control loops can corporate quite well by introducing only sampling interval's delay.

The system dynamics of the paperboard machine can be described by

$$A(z^{-1})y(t) = B(z^{-1})z^{-k}u(t) + C(z^{-1})e(t)$$

where $y(t) = (y_1(t)\ y_2(t))^T$, $u(t) = (u_1(t)\ u_2(t))^T$ and

$$A(z^{-1}) = I + \begin{pmatrix} -1.15 & 0 \\ 0 & -1.32 \end{pmatrix} z^{-1} + \begin{pmatrix} 0.33 & 0.0 \\ 0.0 & 0.43 \end{pmatrix} z^{-2}$$

$$B(z^{-1}) = \begin{pmatrix} 6.1 & -3.4 \\ 0.4 & -1.2 \end{pmatrix} + \begin{pmatrix} -3.36 & 2.03 \\ -0.25 & 0.91 \end{pmatrix} z^{-1}$$

$$C(z^{-1}) = I, \quad k = 6$$

The weighting polynomials in the cost function are chosen as

$$P = I \quad R = I \quad Q' = \begin{pmatrix} 10.0 & 0.0 \\ 0.0 & -1.0 \end{pmatrix} - \begin{pmatrix} 10.0 & 0.0 \\ 0.0 & -1.0 \end{pmatrix} z^{-6}$$

Based on the initial parameters of the process model, the initial parameters of controller can be calculated

$$\tilde{E} = I + \begin{pmatrix} 1.15 & 0 \\ 0 & 1.32 \end{pmatrix} z^{-1} + \begin{pmatrix} 1.0 & 0.0 \\ 0.0 & 1.31 \end{pmatrix} z^{-2}$$

$$+ \begin{pmatrix} 0.76 & 0.0 \\ 0.0 & 1.31 \end{pmatrix} z^{-3} + \begin{pmatrix} 0.54 & 0.0 \\ 0.0 & 0.79 \end{pmatrix} z^{-4} + \begin{pmatrix} 0.37 & 0 \\ 0 & 0.79 \end{pmatrix} z^{-5}$$

$$\tilde{F} = \begin{pmatrix} 0.25 & 0 \\ 0 & 0.61 \end{pmatrix} + \begin{pmatrix} -0.12 & 0.0 \\ 0.0 & -0.33 \end{pmatrix} z^{-1}$$

$$\tilde{G} = I + \begin{pmatrix} 6.1 & -3.4 \\ 0.4 & -1.2 \end{pmatrix} + \begin{pmatrix} 3.66 & -1.9 \\ 0.28 & -0.67 \end{pmatrix} z^{-1}$$

$$+ \begin{pmatrix} 2.2 & -1.0 \\ 0.19 & -0.37 \end{pmatrix} z^{-2} + \begin{pmatrix} 1.31 & -0.57 \\ 0.13 & -0.2 \end{pmatrix} z^{-3}$$

$$+ \begin{pmatrix} 0.79 & -0.31 \\ 0.1 & -0.1 \end{pmatrix} z^{-4} + \begin{pmatrix} 0.47 & -0.17 \\ 0.07 & -0.15 \end{pmatrix} z^{-5}$$

$$+ \begin{pmatrix} -1.28 & 0.77 \\ -0.20 & 0.71 \end{pmatrix} z^{-6}$$

In the parameter estimation algorithm (4.100)-(4.102), the values of the forgetting factors are $\lambda_1 = \lambda_2 = 0.985$. The initial values of covariance matrices are $P_1 = P_2 = I_{16 \times 16}$.

The adaptive control strategy for basis weight and moisture content was implemented in an IBM PC-based process computer using the Basic and Assembly languages. The real-time computer control system has been in operating for five years. The system demonstrates a satisfactory performance. Figures 4.8 and 4.9 show the responses of the system when the set point $w = [w_1, \ w_2]^T$ is changed, respectively, from $[360, 7]^T$ to $[365, 7]^T$ and from $[365, 7]^T$ to $[365, 10]^T$ when $t = 100$. The reel basis weight converges to new set-point within 20 sampling intervals (including 6 intervals time-delay). The reel moisture content converges to new set-point within 40 sampling intervals. Figure 4.10 shows the variation of reel basis weight under the control of the adaptive control system and PID regulators. It can be seen that the variation is greatly reduced after implementing the adaptive control system. The standard deviation of basis weight over 20 days is reduced from $14g/m^2$ to $5g/m^2$. The sheet brakes are significantly reduced. The percentage of top quality product is increased.

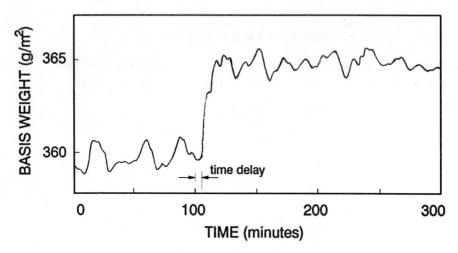

Figure 4.8: Response of basis weight when setpoint changes

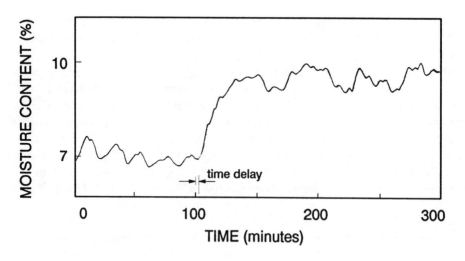

Figure 4.9: Response of moisture content when setpoint changes

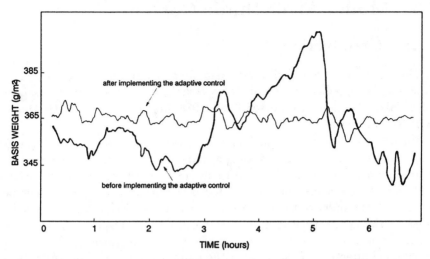

Figure 4.10: Variations of basis weight with and without the adaptive control

Since the paperboard machine produces product with several different grades, we have developed a nominal model for each grade. The model presented in this section is for the most important grade. In grade transition, the system selects the nominal model for the new grade, then starts parameter estimation. To reduce the computation burden of the adaptive algorithm, time delays were not included in on-line parameter estimation. They were estimated off-line for each product grade according to machine speed. It is necessary to point out that if the available process model is accurate enough, the on-line estimation of parameters is not necessary. The parameter estimation is initiated either on the request of operators or when the deviations of basis weight and moisture content to their targets in the past one hour exceed some thresholds.

4.3   Model Algorithmic Control

4.3.1   Introduction

The increasing use of digital minicomputers has resulted in a new control technique, i.e. predictive control. This technique has been proven useful in industrial applications (De Keyser, et al., 1988; Marchetti, et al., 1983; Martin, 1981; Richalet, et al., 1978).

Predictive control is different from the conventional state space and transfer function methods. Both the state space and transfer function methods use parametric models, which is difficult to obtain for most complex multivariable processes. Large errors may be resulted if the order of the model does not match with the order of the real process. The predictive control technique is based on the nonparametric models, namely, impulse response models. The advantage of using impulse response model is that the model coefficients can be obtained directly from samples of the input and output responses without assuming a model order. Moreover, the impulsive response representation is convenient for a constant checking and updating of the model parameters. Other advantages of predictive control technique include inherent time delay compensation and the ability to accommodate process constraints.

Two most popular predictive control techniques are Model Algorithmic Control (MAC) (Cheng, 1989; Rouhani and Mehra, 1982) and Dynamic Matrix Control (DMC) (Culter and Remarker, 1980). Model Algorithmic Control basically involves: (i) an impulse response model for system representation and prediction, (ii) a reference trajectory, (iii) an optimality criterion and, (iv) the consideration of state and control constraints (Marchetti, et al., 1983). The main idea of MAC strategy is to predict the deviation of the future system outputs from the reference path based on the model, to define an optimality criterion which reflect the deviations, and to obtain optimum control strategy to minimize the criterion over a certain horizon in the future (Xia, et al., 1993). The closed-loop MAC, in which process uncertainties can be incorporated by adjusting the discrepancy between the process output and its predicted value, displays a particularly high degree of robustness against process model errors and disturbances.

The paper machine investigated produces white cardboard, which uses a combination of fourdriniers and cylinder vats. The basis weight and moisture content in machine are difficult to control due to the following reasons: (i) the paper machine is of great parameter drifts, external disturbances and random noises; (ii) the dynamics of the machine is nonlinear and time varying; (iii) there exist large time delays in basis weight and moisture content control loops;

(iv) there are serious interactions between the two control loops (Thorp, 1991; Xia, et al., 1993).

This section investigates the application of MAC strategy to basis weight and moisture content control of a cardboard machine. An MAC strategy with feedforward decoupling is presented. An open-loop identification algorithm is also introduced. The effectiveness of the identification algorithm and control algorithm are verified through digital simulations.

4.3.2   Fundamental of the paper machine

The paper machine investigated is a combination of one fourdrinier and six cylinder molds which produces multi-ply cardboard. Figure 4.11 illustrates the layout of the machine.

The four flows of pulp, namely base pulp, middle pulp, filler pulp and top pulp, are pumped into high-level tanks and diluted with white water, respectively. The diluted pulp flows to the distributor, then each of the first three pulp flows is distributed into two. The base pulp, filler pulp and middle pulp flow to six cylinder molds. The top pulp flows to the headbox of the fourdrinier. In the cylinder mold, a horizontal cylinder having a wire cloth surface is arranged to rotate approximately three quarters submerged in a container (vat) of paper stock. Water associated with the fibrous suspension drains through the wire cloth with the result that a layer of fibers is deposited on the surface.

In operation, a moving felt is pressed, by means of roll, into contact with the cylinder at approximately the top position. By doing this, the layer of fibers that has formed on the wire screen is transferred to the mold felt which moves away from the forming screen with it. The felt and web picked up from the first cylinder proceed to the second cylinder, where they are pressed against the web formed there to pick it up. The felt continues through the cylinder mold section, picking up webs from all the six vats in succession.

The top pulp flows to the headbox and forms web on the moving fourdrinier wire. This web is combined with the webs picked up from the previous cylinder molds to form a thick composite board web. The composite web goes through the press section and then passes around two groups of steam-filled cylinders where the remaining water is removed by evaporation. The paper web of moisture content $6 - 10\%$ then goes to coating section. After that the web is dried by the third group of steam-filled cylinders and passes through calender sizing to form the cardboard product.

The operation of the cardboard machine requires the basis weight and moisture content to meet process specifications while minimizing the energy

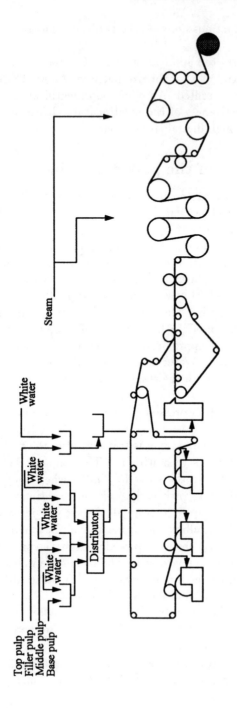

Figure 4.11: Layout of the cardboard machine

and pulp consumption per ton of cardboard product. The basis weight is controlled by adjusting the flow rate of middle pulp, while the moisture content is controlled by adjusting the flow rate of the steam entering the drying section. The difficulties in the cardboard machine control arise from:

(1) The machine produces cardboard of basis weight as heigh as $600g/m^2$. However, the machine speed is low. Greater thickness of the web requires more drying cylinders. This results in large time delay. The time delay from the middle pulp control valve to reel basis weight is about 8 minutes, and that from the steam control valve to reel moisture content is about 2 minutes.

(2) There are serious couplings between the control of basis weight and moisture content.

(3) The capability of the forming cylinders to catch fibers varies with the wear of wires and felts. The varying of machine speed also changes the dynamics of the paperboard machines.

Considering the above problems, we proposed the model algorithmic control strategy with feedforward decoupling for basis weight and moisture content control.

4.3.3   General philosophy of MAC

The principle diagram of the model algorithmic control strategy which is conceptually similar to a model reference adaptive type of control, is depicted in Figure 4.12. MAC strategy involves three parts: the impulse response model, the reference trajectory and the optimality criterion.

Impulse response model

The multivariable process to be controlled is represented by its impulse responses which could be identified both on-line and off-line. However, most of computation time is off-line. This model is used for on-line prediction of future system outputs according to inputs.

MAC strategy makes use of an approximation of the system impulse response g_i by a finite number of terms (N). The model is represented as follows

$$y_m(k+1) = \sum_{i=1}^{N} g_i u(k+1-i) \qquad (4.103)$$

where $y_m(k)$ is the system output at time instant k; $u(k)$ is the system input at time instant k; g_i $(i = 1, 2, \cdots, N)$ are the coefficients of impulse response, which means that we use N terms in the impulse response model and let $g_i = 0$

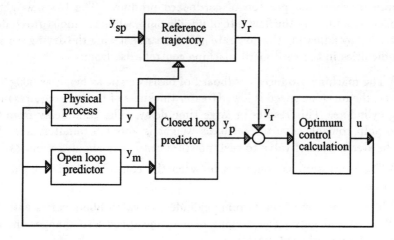

Figure 4.12: Principle diagram of MAC strategy

for $i > N$. $N\Delta T >$ setting time and ΔT is the sampling period.

The reference trajectory

The purpose of MAC is to lead the output $y(t)$ along a desired, and generally smooth path to an ultimate setpoint y_{sp}. The most commonly used reference trajectory is a first order path which initiates on the system output

$$y_r(k+i) = \alpha^i y(k) + (1 - \alpha^i)y_{sp} \qquad (i = 1, 2, \cdots) \qquad (4.104)$$

with

$$\alpha = exp(-T/T_r) \qquad (0 \leq \alpha < 1) \qquad (4.105)$$

where y_{sp} is the setpoint; y_r is the reference trajectory; T_r is the time constant of the reference trajectory. α can be a tuning factor of the controller performance.

Optimality criterion and optimum control strategy

The optimum control strategy is to find a set of future control variables such that the future system output will be as close to the reference trajectory as possible. Thus the optimal control strategy is defined as the one that minimizes the deviation of the predicted outputs from the reference path over a certain horizon in the future. The optimality criterion is defined as

$$J_p = \sum_{i=1}^{P}(y_p(k+i) - y_r(k+1))^2 q_i + \sum_{i=1}^{M} r_i u^2(k-1+i) \qquad (4.106)$$

where P $(\leq N)$ is the horizon of optimization, or horizon of prediction; M $(\leq P)$ is the horizon of the control action allowed to be changed; q_i $(i = 1, 2, \cdots, P)$ and r_i $(i = 1, 2, \cdots, M)$ are nonnegative weighting factors.

Algorithm in the presence of time delay

Most industrial processes, such as paper-making process, have large time delay. For the control of such kind of processes, the MAC algorithm given in equations (4.103)-(4.106) must be modified to incorporate the time delay.

Let

$$\tau = LT_s \qquad (4.107)$$

where τ is time delay, T_s is sampling period, then g_1, g_2, $\cdots$, $g_L = 0$. The modified algorithm is as follows:

Open-loop prediction model:

$$y_m(k+1) = \sum_{i=1}^{N} g_i u(k+1-i) = \sum_{i=L+1}^{N} g_i u(k+1-i) \qquad (4.108)$$

Denoting $k^* = k - L$, i.e. $k = k^* + L$, gives

$$y_m(k^* + L + 1) = \sum_{i=1}^{N-L} g_{L+i} u(k^* + 1 - i) \qquad (4.109)$$

Reference trajectory:

The reference trajectory should be delayed for L sampling intervals because of the presence of the process time delay

$$y_r(k^* + L + i) = \alpha^i y(k^*) + (1 - \alpha^i)y_{sp} \qquad (i = 1, 2, \cdots, P) \qquad (4.110)$$

Closed-loop prediction:

In order to deal with unknown load disturbances, the open-loop predicted values given in (4.108) are further adjusted to the measured actual output $y(t)$ of the system

$$y_p(k^* + L + 1) = y_m(k^* + L + 1) + e(k^*) \qquad (4.111)$$

where

$$e(k^*) = y(k^*) - y_m(k^*) \qquad (4.112)$$

The optimality criterion:

$$J_p = \sum_{i=1}^{P}(y_p(k+L+i) - y_r(k+L+i))^2 q_i + \sum_{i=1}^{M} u^2(k^* + i - 1)r_i \quad (4.113)$$

Since no constraints on outputs and inputs are imposed, the optimum control can be easily obtained from equations (4.109)-(4.113).

4.3.4 Identification algorithm

It is well known that certain restrictions must be imposed on the input signal to guarantee the accuracy of the identification results. In the closed-loop identification, the input is determined through output feedback. The feedback may bring some difficulties in the identification, such as identifiability of the process, bias and inconsistency of the estimate. Actually, for the system without external exciting signal, it is very difficult to obtain accurate identification results.

To avoid this problem, it is very important to introduce external exciting signals. It has been proven that the errors in the identification results are related to the ration of the magnitudes of introduced external signals and process noises. The larger the ratio, the smaller the identification errors.

Because of these reasons, identification is usually conducted on open-loop or very weak closed-loop system with strong external exciting signals. Closed loop identification is usually applied to the open loop unstable system for which it is impossible to obtain useful information but very short data series without a stabilizing feedback. Also safety requirements can be strong reasons for using a regulator in a feedback loop during the identification.

Paper-making processes are open loop stable. Safety is not a major concern during the identification. Therefore, we investigate the open loop identification in this section.

Identification under the concept of minimum structure distance

A definite system can be represented by the following two linear equations

$$y(k) = \mathbf{g}^T \mathbf{u}(k) \qquad (4.114)$$

$$y_m(k) = \mathbf{g}_m^T(k)\mathbf{u}(k) \qquad (4.115)$$

with

$$\mathbf{g} = (g_1 \ g_2 \ \cdots \ g_N)^T$$
$$\mathbf{g}_m = (g_{m1} \ g_{m2} \ \cdots \ g_{mN})^T$$

$$\mathbf{u}(k) = (u(k-1)\; u(k-2)\; \cdots\; u(k-N))^T$$

where $y(k)$ and $y_m(k)$ are the outputs of physical system and its model; $\mathbf{u}(k)$ is control vector; $\mathbf{g}^T$ is the coefficient vector of the system impulse responses, and $\mathbf{g}_m^T(k)$ is that of the model at time instant k.

The errors between the outputs of the model and the physical process are obtained from equations (4.114) and (4.115)

$$e(k) = y_m(k) - y(k) = (\mathbf{g}_m^T(k) - \mathbf{g}^T)\mathbf{u}(k) \tag{4.116}$$

The model distance in the parameter space is defined as

$$D(k) = \parallel \mathbf{g}_m(k) - \mathbf{g} \parallel_Q = (\mathbf{g}_m^T(k) - \mathbf{g}^T)Q(\mathbf{g}_m(k) - \mathbf{g}) \tag{4.117}$$

where Q is a positive definite matrix.

If the following condition is satisfied

$$D(k+1) - D(k) < 0 \tag{4.118}$$

the identification algorithm is convergent.

Defining

$$\Delta\mathbf{g}_m(k) = \mathbf{g}_m(k) - \mathbf{g} \tag{4.119}$$

and

$$
\begin{aligned}
\Delta\mathbf{g}_m(k+1) &= \mathbf{g}_m(k+1) - \mathbf{g} \\
&= \Delta\mathbf{g}_m(k) + \delta\mathbf{g}(k+1)
\end{aligned}
\tag{4.120}
$$

where

$$\delta\mathbf{g}(k+1) = \mathbf{g}_m(k+1) - \mathbf{g}_m(k) \tag{4.121}$$

we obtain

$$D(k+1) - D(k) = 2\Delta\mathbf{g}_m^T(k)Q\delta\mathbf{g}(k+1) + \delta\mathbf{g}^T(k+1)Q\delta\mathbf{g}(k+1) \tag{4.122}$$

and

$$e(k) = \Delta\mathbf{g}_m^T(k)\mathbf{u}(k) \tag{4.123}$$

The identification algorithm is based on the selection of $\delta\mathbf{g}(k+1)$. It is obvious that $\delta\mathbf{g}(k+1)$ is a function of the input vector. We choose

$$\delta\mathbf{g}(k+1) = \nu Q^{-1}\mathbf{u}(k) \tag{4.124}$$

where ν is a real number. From (4.122) and (4.123), it gives

$$D(k+1) - D(k) = \nu^2\mathbf{u}^T(k)Q^{-1}\mathbf{u}(k) + 2\nu e(k) \tag{4.125}$$

To ensure the convergence of the algorithm, we can choose

$$\nu = -\frac{\lambda e(k)}{\mathbf{u}^T(k)Q^{-1}\mathbf{u}(k)} \qquad 0 < \lambda < 2 \tag{4.126}$$

The identification algorithm is then obtained as follows

$$\mathbf{g}_m(k+1) = \mathbf{g}_m(k) - \lambda\frac{e(k)Q^{-1}\mathbf{u}(k)}{\mathbf{u}^T(k)Q^{-1}\mathbf{u}(k)} \tag{4.127}$$

If we choose Q in equation (4.117) to be identity matrix, i.e. define the model distance in the parameter space as

$$D(k) = \| \mathbf{g}_m(k) - \mathbf{g} \| \tag{4.128}$$

the identification algorithm can be simplified into

$$\mathbf{g}_m(k+1) = \mathbf{g}_m(k) - \lambda\frac{e(k)\mathbf{u}(k)}{\mathbf{u}^T(k)\mathbf{u}(k)} \tag{4.129}$$

Simulation of the identification algorithm

The transfer function from the reel basis weight to middle pulp control valve is used as a simulation model. The transfer function is

$$G(s) = \frac{10e^{-8s}}{(1.5s+1)(0.8s+1)} \tag{4.130}$$

The sampling period is $T_s = 0.5$.

The external signal is chosen to be white noise of magnitude c_1, and the process noise to be white noise of magnitude c_2. c_1/c_2 is signal-to-noise ratio. Defining

$$\omega = \sum_{i=1}^{N}(\mathbf{g}_m(i) - \mathbf{g})^2 \tag{4.131}$$

the simulation results are presented in the following table

Table 4.1 Simulation results

c_1/c_2	∞	10	1
ω	0.009	0.020	1.283

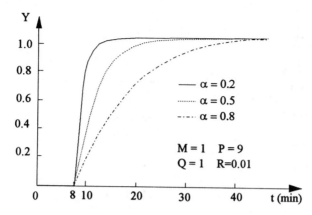

Figure 4.13: Effect of α on system response

We can conclude from the table that the identification algorithm works quite well in the case that the signal-to-noise ratio is much larger than 1. However, when the ratio is close to 1, that is the exciting signal and process noise is of comparative magnitude, the identification result is not accurate enough. This algorithm takes long time (128 minutes) to converge to the acceptable result.

4.3.5   Simulation of MAC for SISO systems

A computer simulation study is performed to investigate the effect of the different design parameters on the performance of the control system. We still use the transfer function given in equation (4.130) as a simulation model. The sampling period is $T_s = 0.5$min, and the number of terms of the impulse responses is $N = 50$.

The main design parameters of the control algorithm are the time constant of the reference trajectory T_r which determines the parameter α in the reference trajectory, the horizon of optimization P and the horizon of the control M. These parameters define the desired behaviour and the stability of the controlled variables. Usually in industrial processes, the goal is to accelerate the natural response of the system within the limitations of the constraints on the actuators.

The simulation results under different control parameters α, P, and M are presented in Figures 4.13-18.

Figure 4.13 compares the performance of the MAC strategy under different α for unit step change in the setpoint. It can be seen that α affects on the

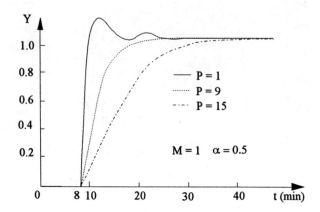

Figure 4.14: Effect of P on system response

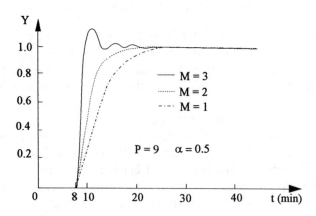

Figure 4.15: Effect of M on system response

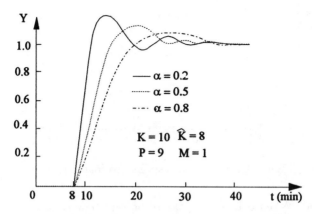

Figure 4.16: Effect of process gain discrepancy under different α

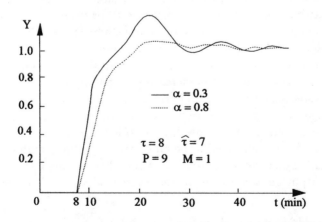

Figure 4.17: Effect of time delay discrepancy under different α

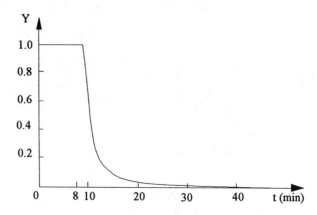

Figure 4.18: System response to load change

response speed. When α increases, the response becomes more sluggish.

The effect of the design parameters P and M on control system performance is shown in Figures 4.14 and 4.15 for the step changes in setpoint. These figures indicate that P and M have the different effect on the performance of the control system. When P increases, the system response slows down and the overshoot decreases. That is, increasing P will improve the stability of the system. Whereas when M increases, the closed-loop responses tend to become more oscillatory and to have larger overshoot. It can also be shown that increasing the weighting factor R (or decreasing Q) will result in more conservative control actions and more stable system. The simulation result for different R and Q are not presented here.

The effect of model errors on the control performance is shown on Figures 4.16 and 4.17. It is indicated that increasing α can improve the robustness of the control system against the model errors. However, the use of larger α will result in a sluggish response. Therefore, we must make a compromise between the robustness and the response speed. Since α has significant effect on system performance, it is recommended to be an on-site adjusting parameter.

Figure 4.18 shows the load response of the MAC strategy. It is seen that the system output quickly returns to the desired value after it is disturbed.

4.3.6   Feedforward decoupling MAC in paper machine

The basis weight and moisture content control system of the cardboard machine investigated is a multivariable coupling systems. We apply feedforward decoupling strategy to decompose the multivariable system into two separated

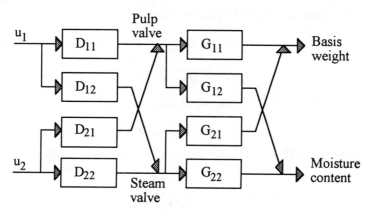

Figure 4.19: Block diagram for forward decoupling design

SISO systems, and then apply the SISO MAC strategy.

The process model of the paper machine is given by

$$G_{11}(s) = \frac{10e^{-8s}}{(1.5s+1)(0.8s+1)} \tag{4.132}$$

$$G_{12}(s) = \frac{0.85e^{-8s}}{(1.5s+1)(0.8s+1)} \tag{4.133}$$

$$G_{21}(s) = \frac{1.25e^{-2s}}{(1.6s+1)(0.1s+1)} \tag{4.134}$$

$$G_{22} = \frac{4.05e^{-2s}}{(1.6s+1)(0.1s+1)} \tag{4.135}$$

The block diagram of the feedforward decoupling MAC system is depicted in Figure 4.19, where D_{ij} $(i,j = 1,2)$ is the feedforward decoupling elements designed to eliminate the interactions between two SISO control systems.

The transfer function relating $Y(s)$ to $U(s)$ is given as

$$\frac{Y(s)}{U(s)} = \begin{pmatrix} D_{11} & D_{12} \\ D_{21} & D_{22} \end{pmatrix} \begin{pmatrix} G_{11} & G_{12} \\ G_{21} & G_{22} \end{pmatrix}$$

$$= \begin{pmatrix} D_{11}G_{11} + D_{12}G_{21} & D_{11}G_{12} + D_{12}G_{22} \\ D_{21}G_{11} + D_{12}G_{21} & D_{21}G_{12} + D_{22}G_{22} \end{pmatrix} \tag{4.136}$$

We specify the transfer functions of the decoupled system to be the same as that of the main paths of the process, that is

$$D_{11}G_{11} + D_{12}G_{21} = G_{11} \tag{4.137}$$

$$D_{21}G_{12} + D_{22}G_{22} = G_{22} \tag{4.138}$$

$$D_{11}G_{12} + D_{12}G_{22} = 0 \tag{4.139}$$

$$D_{21}G_{11} + D_{22}G_{21} = 0 \tag{4.140}$$

From equations (4.137)-(4.140), we obtain the decoupling elements

$$D_{11} = D_{22} = \frac{G_{11}G_{22}}{G_{11}G_{22} - G_{12}G_{21}} \tag{4.141}$$

$$D_{12} = -\frac{G_{11}G_{12}}{G_{11}G_{22} - G_{12}G_{21}} \tag{4.142}$$

$$D_{21} = -\frac{G_{22}G_{21}}{G_{11}G_{22} - G_{12}G_{21}} \tag{4.143}$$

Substituting the process model into the above equations gives

$$D_{11} = D_{22} = 1.03 \quad D_{12} = -0.21e^{-6s} \quad D_{21} = -0.15e^{6s}$$

The decoupling element D_{21} is not realizable because of the presence of the pure lead part e^{6s}. In the implementation of the feedforward decoupling control, we drop the pure lead term in D_{21}.

It is found in the simulation study that such a decoupling strategy brings some disadvantages:

(1) Since the time delays in the two control loops are quite different, the decoupling component D_{21} is not realizable. If we simply drop the pure lead term, the decoupling performance is poor.

(2) Even if we assume that the time delay of the moisture content loop is the same as that of the basis weight loop, the decoupling performance is still not satisfactory. This problem is caused by the fact that the sample period $T_s = 0.5min$ is too large compared with the time constant of the process. The zero-order holder is not able to make the two control loops cooperate well, and thus results in the poor decoupling performance.

In order to solve the problems above, we apply the following modifications in the system implementation:

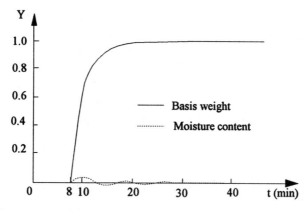

Figure 4.20: Response for unit step change of basis weight

(1) After the control action for steam control valve is obtained, it is hold for $6min$ (12 sampling intervals) before issued.

(2) When the decoupling elements G_{12} and G_{21} are discretized, an additional term $\lambda e^{\mu T_s}$ is also included together with the zero-order holder. The purpose of introducing the addition term is for gain and phase compensation of the discrete system. Satisfactory decoupling performance is obtained by adjusting the values of λ and μ.

The simulation results of the MAC system by using the above feedforward decoupling are presented in Figures 4.20 and 4.21. It indicates that the closed loop system has satisfactory decoupling and regulatory performance. The effect of the basis weight change on the moisture content is 0.2% and that of the moisture content on the basis weight is 6.2%.

The basis weight of the different grades of the cardboard product varies from $250g/m^2$ to $600g/m^2$. It is impossible to use a single model to represent the process dynamics for different grades of product. To solve the problem, we divide the basis wight into three regions, $250 - 350g/m^2$, $350 - 450g/m^2$ and $450 - 600g/m^2$. For each regions, a linear process model is developed. The control system is designed based on each of these models, respectively. The system selects the different control parameters according to product grades (basis weight specifications). This multi-model technique has a low computation burden. However, satisfactory control performance can be reached.

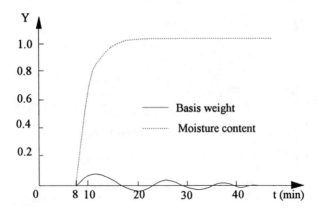

Figure 4.21: Response for unit step change of moisture content

4.4   Conclusions

This chapter presented three algorithms to solve the problems mainly coming from unmeasurable states and large time delays. An adaptive fading Kalman filter algorithm for state estimation was presented. The algorithm improved both optimality and convergence. The filter used variable exponential weighting method to compensate the model errors and unknown drifts. Since there was only one adjustable factor in the algorithm, complete optimality might not be ensured. In the case where higher degree of optimality is required, matrix forgetting factor should be considered instead of scalar factor in order to provide different rates of fading for different filter channels.

A successful application of adaptive control strategy in a paperboard machine was also presented. The combination of conventional controls, a multivariable k-incremental predictor, a multivariable self-tuning controller had been shown to be very effective on the control of the basis weight and moisture content on a paperboard machine. A measure of the success and acceptance of the new control system is that an industrial paperboard machine has been controlled by this strategy for five years and now are still in operation. The significant reduction in the variation of basis weight and moisture content implies good economic return.

Model Algorithmic Control strategy (MAC) is very promising for the process with large time delay. It has the advantages of convenience in implementation and good robustness to model errors and external disturbances. Its application of a cardboard machine was discussed. Feedforward decoupling was applied to eliminate the interaction between the basis weight and moisture content control loops. The performance of the identification algorithm was tested under the different signal-to-noise ratios. The simulation results indicated that developed feedforward decoupling MAC strategy was very effective on the control of the basis weight and moisture content.

4.5   References

Allison B.J., Dumont G.A. and Novak L.H. Multi-input adaptive-predictive control of kamyr digester chip level. Canadian J. of Chem. Eng. 1991; 69:111-119

Åström K.J. Computer control of a paper machine – an application of linear stochastic control theory. IBM J. Res. Dev. 1967; 11:389-405

Åström K.J., Borisson U., Ljung L. and Wittenmark B. Theory and application of self-tuning regulators. Automatica 1977; 13:457-476

Bialkowski W. L. Applications of Kalman filters to the regulation of dead time processes. IEEE Trans. Autom. Contr. 1983; AC-28:400-406

Borrison U. Self-tuning regulators: industrial application and multivariable theory. Report 7513, Department of Automatic Control, Lund Institute of Technology, 1975

Cegrell T. and Hedquist T. Successful adaptive control of paper machines. Automatica 1975; 11:53-59

Chang C.B. and Tabaczynski J.A. Application of state estimation to target tracking. IEEE Trans. Autom. Contr. 1984; AC-29:98-109

Cheng C.M. Linear quadratic-model algorithmic control method: a controller design method combining the linear quadratic model algorithmic control algorithm. Industrial & Engineering Chemistry Research 1989; 28:178-186

Clarke D.W., Hodgson A.J.F. and Tuffs P.S. Offset problem and k-incremental predictors in self-tuning control. IEE Proc 1983; 130-D:217-225

Culter G.R. and Remarker B.L. Dynamic matrix control – A computer control algorithm. JACC, WP5-B, San Francisco, 1980

De Keyser R.M.C., Van de Velde P.G.A. and Dumortier F.A.G. Comparative study of self-adaptive long-range predictive control methods. Automatica 1988; 24:149-163

Deyst J.J.Jr. and Price C.F. Conditions for asymptotic stability of the discrete minimum-variance linear estimator. IEEE Trans. Autom. Contr. 1968; AC-13:702-705

Fagin S.L. Recursive linear regression theory, optimal filter theory, and error analysis of optimal system. IEEE International Convention Record 1964; 12:216-240

Fitzgerald R.L. Divergence of the Kalman filter. IEEE Trans. Autom. Contr. 1971; AC-16:736-747

Kalman R.E. and Bucy R.S. New results in linear filtering and prediction theory. Trans. ASME J. Basic Engng 1961; 83:95-108

Koivo H.N. and Tanttu T.J. Self-tuning controllers: non-square systems and convergence. Int. J. System Sci. 1985; 16:981-1002

Marchetti J.L, Mellichamp D.A. and Seborg D.E. Predictive Control Based on the Discrete Convolution Model. I & EC Process Des. Dev. 1983; 22:488-495

Martin G.D. Long range predictive control. J. AIChE 1981; 27:748-753

Masreliez C.J. and Martin R.D. Robust Bayesian estimation for the linear model and robustifying the Kalman filter. IEEE Trans. Autom. Contr. 1977; AC-22:361-371

Maybeck P.S. Stochastic models, estimation, and control. Vol. 1-3, Academic Press New York, 1982

Mayne D.Q., Åström K.J. and Clark J.M.C. A new algorithm for recursive estimation of parameters in controlled ARMA processes. Automatica 1984; 20:751-760

Ohap R.H. and Stubberud A.R. Adaptive minimum variance estimation in discrete time linear systems. In: Control and Dynamic Systems, C.T. Leondes, editor, Vol 12. Academic Press, New York, 1976, pp 583-642

Polak E. Computational methods in optimization; a unified approach. Series in Mathematics in Science and Engineering, Vol.77, Academic Press, New York, 1971

Richalet J., Rault A., Testud J.L. and Papon J. Model predictive heuristic control: applications to industrial processes. Automatica 1978; 14:413-428

Rouhani R. and Mehra R.K. Model algorithmic control (MAC): basic theoretical properties. Automatica 1982; 18:401-414

Shead R.P. Computer control improves coated boxboard machine productivity. Pulp and Paper 1979; 53(2):130-133

Shellenbarger J.C. Estimation of convariance parameter for an adaptive Kalman filter. Proc National Electronics Conference, 1966, pp 698-702

Sikora R.F. and Bialkowski W.L. A self-tuning strategy for moisture control in paper making. Proc American Control Conference, San Diego, California, 1984, pp 54-60

Sorenson H.W. and Sacks J.E. Recursive fading memory filtering. Information Sciences 1971; 3:101-119

Synder D.L. Information processing for observed jump processes. Information and Control 1973; 22:69-75

Thorp B.A. (editor) Paper Machine Operations. In: Pulp and Paper Manufacture (3rd ed.), Vol.7, 1991

Tsai C. and Kurz L. An adaptive robustizing approach to Kalman filtering. Automatica 1983; 19:279-288

Tung L.S. Sequential predictive control of industrial processes. Proc American Control Conference, San Francisco, California, 1983, pp 349-355

Tze-Kwarfung P. and Grimble M.J. Dynamic ship positioning using a self-tuning Kalman filter. IEEE Trans. Autom. Contr. 1983; AC-28:339-350

Wallace B.W. Process control systems: harnessing run away energy. PIMA 1980; 62:22-26

Wilhelm R.G. Self-tuning strategies: multi-faceted solution to paper machine control problem. Presented at ISA Spring Symp., Columbus, Ohio, 1982

Xia Q. Reliable control theory and application in industrial processes. Ph.D. dissertation, Zhejiang University, Hangzhou, PRC, 1989

Xia Q., Rao M., Shen X. and Zhu H. Adaptive control of a paperboard machine. Pulp & Paper Canada 1994; (in press)

Xia Q., Rao M., Ying Y. and Shen X. Adaptive fading Kalman with applications. Automatica 1994; (in press)

Xia Q., Rao M. and Qian J. Model algorithmic control of paper machines. Proc of the Second IEEE Conference on Control Applications, Vacouver, Canada, 1993, pp 203-208

Ydstie B.E. and Co T. Recursive estimation with adaptive divergence control. IEE Proceedings 1985; 132D:124-130

CHAPTER 5
BILINEAR CONTROL

In this chapter, the bilinear control strategy for paper machines is discussed based on the characteristics of paper-making process. The bilinear decoupling control, bilinear observer and bilinear optimal control as the typical techniques are discussed. Their applications to the headbox section and drying section of a paper machine, are presented. Digital simulation and on-site implementation results show that the performance of the bilinear control system is satisfactory.

5.1 Introduction

As we know, in the paper machine control system, there exist nonlinear property and interaction among the process variables which will affect the system performance if they are not controlled properly. Also, some process variables cannot be measured by on-line sensors. Therefore, the nonlinear multivariable decoupling control, optimal control and state estimation techniques will be very important in the paper-making process control applications.

Bilinear systems (BLS), as the special nonlinear systems, have attracted much attention in recent years (Ying, et al., 1992). They are linear individually with respect to the state and the control variables, but not jointly. They can be characterized by the following dynamic equation:

$$\dot{X}(t) = AX(t) + BU(t) + \{NX(t)\}U(t) \tag{5.1}$$

where $X \in R^n$, $U \in R^m$, are the state and input (control) vectors, respectively. $\{NX(t)\}U(t)$ is a bilinear form in the variables $X(t)$ and $U(t)$, which in a less concise form may be written as:

$$\{NX(t)\}U(t) = \sum_{i=1}^{m} N_i X(t) U_i(t) \tag{5.2}$$

123

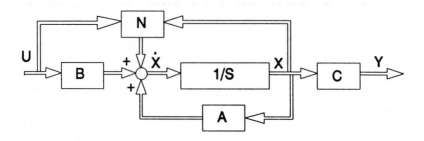

Figure 5.1: Block diagram of the bilinear system

where, $U_i(t)$ is ith component of $U(t)$. The block diagram of BLS is shown in Figure 5.1.

BLS has been applied to many disciplines, such as socioeconomic, ecology, agriculture, biology, and industrial processes. The significant progress on its modelling and control has been achieved (Mohler, 1973; Bruni, et al., 1974; Mohler and Kolodziej, 1980). The research results obtained show that the bilinear systems are generally more controllable and offer better performance than linear systems.

In this chapter, the decoupling control for bilinear multivariable systems (BLMS) is discussed in Section 5.2. Like linear systems, the state and disturbance observers for BLS must be considered in order to realize the state feedback and/or disturbance feedforward control in the case where some of the states and/or disturbances are unmeasurable. Section 5.3 addresses the design issue of minimal order state observers for BLMS; then a new method to design minimal order state-disturbance composite observers is discussed. The applications of the state-disturbance composite observer and the feedback decoupling controller to the headbox control system in paper-making process are presented.

In Section 5.4, suboptimal control for BLS is discussed by using an extension of the linear-quadratic optimal control index. The design method of this bilinear suboptimal control system and its application to the moisture control of a paper machine are also described.

5.2   Bilinear Decoupling Control

The decoupling control for linear multivariable systems has been well developed and successfully applied to the real world process control. For nonlinear systems, although a significant progress on the theoretical research has been achieved (Freund, 1973, 1975; Isidori, 1985), difficulty still remains in real industrial processes. In this section, we will discuss bilinear decoupling control and propose a new design method using state feedback decoupling control for BLMS.

5.2.1   Decoupling controller for bilinear systems

Consider a BLMS described by the following equations

$$\dot{X}(t) = AX(t) + B_0U(t) + \sum_{i=1}^{m} B_iX(t)U_i(t) \qquad (5.3)$$

$$Y(t) = CX(t) + D_0U(t) + \sum_{i=1}^{m} D_iX(t)U_i(t) \qquad (5.4)$$

where $X \in R^n$, $U \in R^m$, $Y \in R^m$ are state vector, input vector, and output vector respectively, U_i is the ith element of vector U, A, B_0, B_i, C, D_0, and D_i $(i = 1, 2, ...m)$ are system coefficient matrices with appropriate dimensions. Let us define

$$B(X,t) = (B_{01} + B_1X(t) \quad B_{02} + B_2X(t) \quad ... \quad B_{0m} + B_mX(t)) \qquad (5.5)$$

$$D(X,t) = (D_{01} + D_1X(t) \quad D_{02} + D_2X(t) \quad ... \quad D_{0m} + D_mX(t)) \qquad (5.6)$$

where B_{0i} is the ith column of B_0, D_{0i} is the ith column of D_0. Then, the original system (5.3) and (5.4) under the new notations is

$$\dot{X}(t) = AX(t) + B(X,t)U(t) \qquad (5.7)$$

$$Y(t) = CX(t) + D(X,t)U(t) \qquad (5.8)$$

The design purpose is to find a state feedback control law $U(t)$

$$U(t) = F_dX(t) + GV(t) \qquad (5.9)$$

which can decouple the input-output of system (5.7) and (5.8), and assign the desired poles for the closed-loop system. Here, F_d and G are $m \times n$ and $m \times m$ coefficient matrices respectively, and $V(t)$ is a set-point vector (see Figure 5.2).

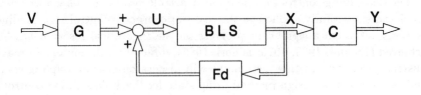

Figure 5.2: Bilinear decoupling control system

Definition *(Nijmeijer and Schaft, 1990): The nonlinear system (5.7) and (5.8) is called input-output decoupled if , after a possible relabeling of the inputs, the following two properties hold:*

(i) For each $i \in m$ the output Y_i is invariant under the inputs V_j, $j \neq i$.

(ii) The output Y_i is not invariant with respect to the input V_i, $i \in m$. where V_i and Y_i are the ith component of $V(t)$ and $Y(t)$, respectively.

Substituting equation (5.9) into equations (5.7) and (5.8) leads to the closed loop system as follows

$$\dot{X}(t) = [A + B(X,t)F_d]X(t) + B(X,t)GV(t) \tag{5.10}$$

$$Y(t) = CX(t) + D(X,t)F_dX(t) + D(X,t)GV(t) \tag{5.11}$$

From equation (5.11) we know that the ith component of the output vector $Y(t)$ is

$$Y_i(t) = C_iX(t) + D_i(X,t)F_dX(t) + D_i(X,t)GV(t) \tag{5.12}$$

where C_i is the ith row of matrix C, and $D_i(X,t)$ is the ith row of the matrix $D(X,t)$. When $D_i(X,t) \neq 0$, the equation (5.12) shows that the input $V_i(t)$ directly affects the output $Y_i(t)$, and the differential order of (5.8) is

$$d_i = 0 \tag{5.13}$$

When $D_i(X,t) = 0$, equation (5.12) becomes

$$Y_i(t) = C_iX(t) \tag{5.14}$$

This means that the input $V_i(t)$ must, through d_i integral links, affect the output $Y_i(t)$, and the differential order d_i of system (5.7) and (5.8) is

$$d_i > 0 \tag{5.15}$$

In region R_0 of the space (X,t), the differential order d_i of the ith output $Y_i(t)$ $(i=1, 2, ..., m)$ for the bilinear systems can be determined as follows:

$$d_i = 0 \qquad \text{for } D_i(X,t) \neq 0$$
$$d_i = \min\{j : |C_i A^{j-1} B(X,t)| \neq 0\} \qquad \text{for } D_i(X,t) = 0$$

Now it is imperative to discover the relationship between the input $V_i(t)$ and the d_ith-order derivative of the output $Y_i(t)$.

Let there be an arbitrary integer d_i for all $(X,t) \in R_0$ for $i = 1, 2, ..., m$, such that $C_i B(X,t)$, $C_i AB(X,t)$, ..., $C_i A^{d_i-2} B(X,t)$ are all null matrices and $C_i A^{d_i-1} B(X,t)$ is a nonsingular matrix (Sinha, 1984). With this assumption, from equation (5.14), we have

$$\dot{Y}_i(t) = C_i \dot{X}(t) = C_i A X(t) \tag{5.16}$$

The 2nd-order derivative of the output can be obtained through differential operation

$$\ddot{Y}_i(t) = C_i A^2 X(t) \tag{5.17}$$

Similarly, the d_ith-order derivative will be

$$\begin{aligned} Y_i^{(d_i)}(t) &= C_i A^{d_i} X(t) + C_i A^{d_i-1} B(X,t) F_d X(t) \\ &+ C_i A^{d_i-1} B(X,t) GV(t) \end{aligned} \tag{5.18}$$

Consider all components of the output vector $Y(t)$, we have

$$Y^*(t) = (C^* + D^* F_d) X(t) + D^* GV(t) \tag{5.19}$$

where, $Y^*(t) = (Y_1^{(d_1)}(t)\ Y_2^{(d_2)}(t)\ ...\ Y_m^{(d_m)})^T$, $C^* \in R^{m \times n}$, $D^* \in R^{m \times m}$, the ith row vector of C^* and D^* are

$$C_i^* = C_i A^{d_i} \qquad (i = 1, 2, ..., m) \tag{5.20}$$

$$D_i^* = D_i(X,t) \qquad d_i = 0 \quad (i = 1, 2, ..., m) \tag{5.21}$$

$$D_i^* = C_i A^{d_i-1} B(X,t) \qquad d_i \neq 0 \quad (i = 1, 2, ..., m) \tag{5.22}$$

According to the definition and the decoupling controller design method for both linear and nonlinear systems (Freund, 1973, 1975; Sinha, 1977; Nazar and Rekasius, 1971; Li and Feng, 1987; Isidor, et al., 1981) we can have the following theorem.

Theorem 1 *For system (5.3) and (5.4), the decoupling controller with the arbitrary pole-assignment for bilinear systems as follows*

$$F_d = -D^{*-1}(C^* + M^*) \tag{5.23}$$

$$G = D^{*-1}\Lambda \tag{5.24}$$

if D^* is non-singular on R_0. Where, M^* is an $m \times n$ matrix, the ith row vector M_i^* $(i = 1, 2, ..., m)$ is

$$M_i^* = \begin{cases} 0 & d_i = 0 \\ \sum_{k=0}^{d_i-1} \alpha_{ik} C_i A^k & d_i \neq 0 \end{cases}$$

Λ *is an $m \times m$ diagonal matrix, where the diagonal elements are $\lambda_i's$. λ_i and α_{ik} $(i = 1, 2, ..., m)$ can be selected arbitrarily in order to get the desired pole-assignment.*

Proof: Substituting equations (5.23) and (5.24) into equation (5.19), we obtain

$$\begin{aligned} Y^*(t) &= \{C^* + D^*[-D^{*-1}(C^* + M^*)]\}X(t) + D^* D^{*-1}\Lambda V(t) \\ &= -M^* X(t) + \Lambda V(t) \end{aligned} \tag{5.25}$$

If $d_i = 0$, then $M^* = 0$, so that

$$Y_i^*(t) = \lambda_i V_i(t) \tag{5.26}$$

If $d_i \neq 0$, from equation (5.25), we have

$$Y_i^{(d_i)}(t) = -(\sum_{k=0}^{d_i-1} \alpha_{ik} C_i A^k)X(t) + \lambda_i V_i(t) \tag{5.27}$$

Since

$$Y_i^{(k)}(t) = C_i A^k X(t) \qquad (k = 0, 1, ..., d_i - 1)$$

then

$$Y_i^{(d_i)}(t) = -\sum_{k=0}^{d_i-1} \alpha_{ik} Y_i^{(k)}(t) + \lambda_i V_i(t)$$

or it can be written as

$$Y_i^{(d_i)}(t) + \alpha_{i,d_i-1} Y_i^{(d_i-1)}(t) + ... + \alpha_{i,0} Y_i(t) = \lambda_i V_i(t) \tag{5.28}$$

Obviously, the poles of the ith decoupling subsystem can be assigned arbitrarily by selecting the appropriate values of $\alpha_{ik}'s$.

$$\text{Q.E.D.}$$

From the discussion above, it is obvious that the necessary condition for the existence of a bilinear decoupling controller is D^* is non-singular on R_0.

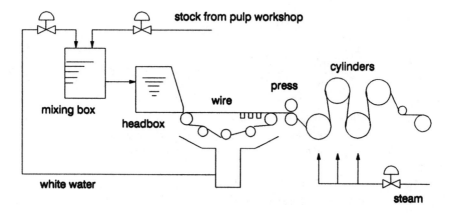

Figure 5.3: Simplified flow chart of a paper-making process

5.2.2   Applications to paper machine headbox

Mathematical model of process

A paper machine that produces super thin condenser paper can be described concisely in Figure 5.3. The stock from pulp preparation is pumped into a mixing box where the stock is mixed with white water, then the mixture is filled into the headbox through a filter in which the dregs in stock are removed. The next step is to place the stock onto the forming wire and to remove most water to form paper. The paper sheet goes through the press part and dryer section (cylinder) to remove the remaining water, and subsequently to accomplish the process of production.

The headbox control system is very important in paper-making process. Since the paper machine which produces super thin condenser paper has a slow machine rate (70m/min), its headbox is usually not pressurized, but open to atmosphere. The level and consistency of stock in the headbox are the key factors affecting production quality. In general, we take the flow rate of white water and stock going into the mixing box as control variables to control the level and consistency of stock in the headbox. Since there exists strong interaction between level control system and consistency control system, the decoupling control strategy is needed in order to improve the control system performance.

The bilinear mathematical model for this headbox system can be formulated as follows.

For the mixing tank, according to the mass balance law, we have

$$\frac{dH_1}{dt} = \frac{1}{A_1}(G_p + G_w - G_1) \tag{5.29}$$

$$\frac{dN_1}{dt} = \frac{1}{A_1 H_1}(G_p N_p + G_w N_w - G_1 N_1) \tag{5.30}$$

where
  G_p, N_p are flow rate and consistency of the stock from pulp workshop,
  G_w, N_w are flow rate and consistency of the white water,
  G_1, N_1 are flow rate and consistency of the mixing stock out of mixing tank,
  A_1, H_1 are the cross-section area and liquid level of the mixing tank,
  Using steady state operating conditions

$$G_{p0} + G_{w0} = G_{10}$$

$$G_{p0} N_{p0} + G_{w0} N_{w0} = G_{10} N_{10}$$

and approximate relationship

$$\Delta G_1 = \Delta H_1 / R_1$$

where R_1 is the flow resistance of mixing tank, and taking deviation of equations (5.29) and (5.30), we get

$$\frac{dH_1}{dt} = \frac{1}{A_1}(G_p + G_w - H_1/R_1) \tag{5.31}$$

$$\frac{dN_1}{dt} = \frac{1}{A_1 H_{10}}[(N_{p0} - N_{10})G_p + (N_{w0} - N_{10})G_w$$
$$+ \quad G_{p0} N_p + G_{w0} N_w - G_{10} N_1 - G_p N_1 - G_w N_1] \tag{5.32}$$

Similarly, for the headbox, we have

$$\frac{dH_2}{dt} = \frac{1}{A_2 R_1} H_1 - \frac{1}{A_2 R_2} H_2 \tag{5.33}$$

$$\frac{dN_2}{dt} = \frac{H_{10}}{A_2 H_{20} R_1} N_1 + \frac{N_{10}}{A_2 H_{20} R_1} H_1$$
$$- \frac{N_{20}}{A_2 H_{20} R_2} H_2 - \frac{1}{A_2 R_2} N_2 \tag{5.34}$$

where
  H_2, N_2 are the level and consistency of the stock in headbox,
  A_2 is the cross-section area of the headbox, and
  R_2 is the flow resistance of the headbox.
Substituting all steady state data,

$$
\begin{array}{ll}
G_{po}=5.82 \text{ T/h} & N_{po}=1.015 \ \% \\
G_{wo}=11.64 \text{ T/h} & N_{wo}=0.05 \ \% \\
H_{10}=650 \ mm\,H_2O & H_{20}=190 \ mm\,H_2O \\
N_{10}=0.35 \ \% & N_{20}=0.34 \ \%
\end{array}
$$

into equations (5.31)-(5.34), we finally get the bilinear model for the headbox system as following

$$\dot{X}(t) = AX(t) + B_0 U(t) + \sum_{i=1}^{2} B_i X(t) U_i(t) + FW(t) \tag{5.35}$$

$$Y(t) = CX(t) \tag{5.36}$$

where
  the state variables $X^T(t) = (X_1(t) \ X_2(t) \ X_3(t) \ X_4(t)) = (H_1 \ H_2 \ N_1 \ N_2)$
  the output variables $Y^T(t) = (Y_1(t) \ Y_2(t)) = (H_2 \ N_2)$,
  the control variables $U^T(t) = (U_1 \ U_2) = (G_p \ G_w)$
  the disturbance variables $W^T(t) = (N_p \ N_w)$

$$
A = \begin{pmatrix}
-1.93 & 0 & 0 & 0 \\
0.394 & -0.426 & 0 & 0 \\
0 & 0 & -0.63 & 0 \\
0.82 & -0.784 & 0.413 & -0.426
\end{pmatrix}
$$

$$
B_0 = \begin{pmatrix}
1.274 & 1.274 \\
0 & 0 \\
1.34 & -0.65 \\
0 & 0
\end{pmatrix}
\quad
B_1 = B_2 = \begin{pmatrix}
0 & 0 & 0 & 0 \\
0 & 0 & 0 & 0 \\
0 & 0 & -0.327 & 0 \\
0 & 0 & 0 & 0
\end{pmatrix}
$$

$$
C = \begin{pmatrix}
0 & 1 & 0 & 0 \\
0 & 0 & 0 & 1
\end{pmatrix}
\quad
F = \begin{pmatrix}
0 & 0 \\
0 & 0 \\
0.203 & 0.406 \\
0 & 0
\end{pmatrix}
$$

Decoupling controller

Based on the bilinear model of the headbox system described above, we have

$$B(x,t) = (B_{01} + B_1 X(t) \quad B_{02} + B_2 X(t))$$

$$D(x,t) = 0$$

Since $d_1 = d_2 = 2$, the state feedback decoupling control (5.9) is

$$U(t) = -D^{*^{-1}}(C^* + M^*)X(t) + D^{*^{-1}} \Lambda V(t) \tag{5.37}$$

in which

$$C^* = \begin{pmatrix} C_1^* \\ C_2^* \end{pmatrix} = \begin{pmatrix} -0.93 & 0.181 & 0 & 0 \\ -2.24 & 0.667 & -0.435 & 0.181 \end{pmatrix} \tag{5.38}$$

$$D^{*^{-1}} = \begin{pmatrix} -1.882 + 0.328x_3(t) & 1.218 \\ 3.876 - 0.328x_3(t) & -1.218 \end{pmatrix} \tag{5.39}$$

while M^* can be obtained as follow

$$M^* = \begin{pmatrix} M_1^* \\ M_2^* \end{pmatrix} = \begin{pmatrix} \sum_{k=0}^{1} \alpha_{1k} C_1 A^k \\ \sum_{k=0}^{1} \alpha_{2k} C_2 A^k \end{pmatrix}$$

$$= \begin{pmatrix} \alpha_{10} C_1 A^0 + \alpha_{11} C_1 A^1 \\ \alpha_{20} C_2 A^0 + \alpha_{21} C_2 A^1 \end{pmatrix} \tag{5.40}$$

In compliance with the control system requirement, the poles of the closed loop system are assigned as

$$s_{11} = -1, \quad s_{21} = -2, \quad s_{12} = -1, \quad s_{22} = -2$$

which implies that

$$\alpha_{10} = 2, \quad \alpha_{20} = 2, \quad \alpha_{11} = 3, \quad \alpha_{21} = 3$$

thus

$$M^* = \begin{pmatrix} M_1^* \\ M_2^* \end{pmatrix} = \begin{pmatrix} 1.18 & 0.722 & 0 & 0 \\ 2.46 & -2.351 & 1.239 & 0.722 \end{pmatrix} \tag{5.41}$$

Taking Λ as a diagonal matrix

$$\Lambda = \begin{pmatrix} 2 & 0 \\ 0 & 3 \end{pmatrix}$$

Finally we obtain

$$F_d = -D^{*-1}(C^* + M^*)$$

$$= \begin{pmatrix} 0.211 - 0.082x_3(t) & 3.75 - 0.296x_3(t) & -0.98 & -1.10 \\ -0.702 + 0.082x_3(t) & -5.55 + 0.296x_3(t) & 0.98 & 1.10 \end{pmatrix}$$

$$(5.42)$$

$$G = D^{*-1}\Lambda$$

$$= 2\begin{pmatrix} -1.88 + 0.328x_3(t) & 1.218 \\ 3.876 - 0.329x_3(t) & -1.218 \end{pmatrix} \qquad (5.43)$$

Simulation and on-line operation performance

The simulation results for this bilinear decoupling control system are shown in Figures 5.4 and 5.5, which indicate that this bilinear multivariable decoupling control system performs very well. When the setpoint of stock consistency has a unit step change, the controlled variable (consistency of stock) can reach the setpoint quickly, while the liquid level remains unchanged. Similarly, when the setpoint of liquid level has a unit step change, the state of the level can reach its setpoint very fast, while the state of the stock consistency keeps constant.

The on-line operation results are shown in Figures 5.6 and 5.7. Due to the model error, the system cannot be completely decoupled, however it can satisfy the production requirements.

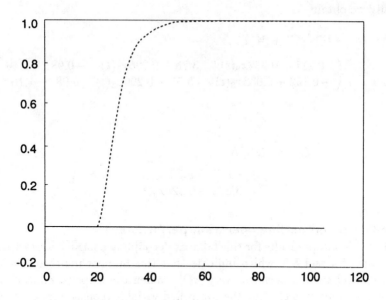

Figure 5.4: Simulation result for unit step change in the setpoint of H_2

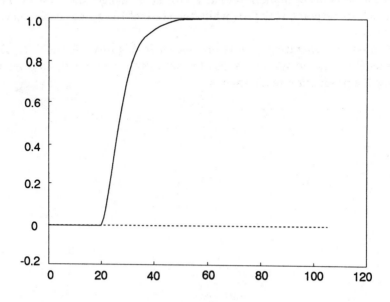

Figure 5.5: Simulation result for unit step change in the setpoint of N_2

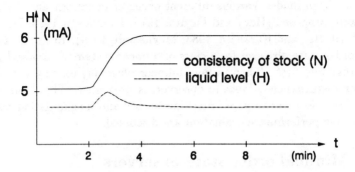

Figure 5.6: Control system responses for unit step change in setpoint of N_2

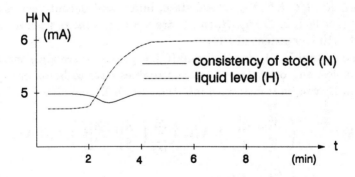

Figure 5.7: Control system responses for unit step change in setpoint of H_2

5.3   Bilinear State Observers

During the last decade, significant progress on observer design for bilinear systems has been made. Various different observer equations and design procedures were proposed (Hara and Furuta, 1976; Funahashi, 1978; Derese and Noldus, 1979; Hsu and Karanam, 1983; Maghsoodi, 1989). In this section, the minimal order state observer for a class of bilinear systems (Hara and Furuta, 1976) is reviewed first. Then, a new bilinear observer, namely the minimal order state-disturbance composite observer, is presented. The application of this composite observer to paper machine control and its simulation results as well as on-line performance evaluation are discussed.

5.3.1   Minimal order state observers

Consider an n-dimensional bilinear system without uncontrollable states, denoted as $\Sigma_p(A^i, B, C)$

$$\dot{X}(t) = A^0 X(t) + \sum_{i=1}^{m} A^i U_i(t) X(t) + BU(t) \tag{5.44}$$

$$X(0) = X_0$$

$$Y(t) = CX(t) \tag{5.45}$$

where, $X \in R^n$, $U \in R^m$, $Y \in R^p$ are state, input, and output vectors respectively; A^i $(i = 0, 1, 2, ..., m)$, B, and C are $n \times n$, $n \times m$, and $p \times n$ matrices. $U_i(t)$ is the ith component of $U(t)$.

For the system $\Sigma_p(A^i, B, C)$, if $\text{rank}(C) = p$, (i.e., there are p measurable state variables, and other $(n - p)$ state variables have to be observed), then the system Σ_p can be transformed into the system Σ_p^* as follows

$$\dot{X}(t) = \begin{pmatrix} A_{11}^0 & A_{12}^0 \\ A_{21}^0 & A_{22}^0 \end{pmatrix} X(t) + \sum_{i=1}^{m} \begin{pmatrix} A_{11}^i & A_{12}^i \\ A_{21}^i & A_{22}^i \end{pmatrix} U_i(t) X(t)$$

$$+ \begin{pmatrix} B_1 \\ B_2 \end{pmatrix} U(t) \tag{5.46}$$

$$Y(t) = \begin{pmatrix} I_p & 0 \end{pmatrix} X(t) \tag{5.47}$$

where, $A_{11}^i \in R^{p \times p}$, $A_{12}^i \in R^{p \times (n-p)}$, $A_{21}^i \in R^{(n-p) \times p}$, $A_{22}^i \in R^{(n-p) \times (n-p)}$, $(i = 0, 1, ..., m)$, $B_1 \in R^{p \times m}$, $B_2 \in R^{(n-p) \times m}$.

For the system Σ_p^*, we have the following theorem (Hara, et al., 1976).

Theorem 2 *For the system Σ_p^*, if there exists an $(n\text{-}p)\times p$ matrix H that satisfies the following two conditions,*

(1) $A_{22}^i + HA_{12}^i = 0 \qquad (i=1,2,...,m)$

(2) All eigenvalues of $(A_{22}^0 + HA_{12}^0)$ have negative real parts,

then there must exist an $(n\text{-}p)$ order (Minimal order) state observer Σ_0, which can be represented as

$$\dot{Z}(t) = \hat{A}^0 Z(t) + \sum_{i=1}^{m} \hat{A}^i U_i(t) Z(t) + \hat{B}^0 Y(t)$$

$$+ \sum_{i=1}^{m} \hat{B}^i U_i(t) Y(t) + \hat{J} U(t) \tag{5.48}$$

$$W(t) = \hat{C} Z(t) + \widehat{D} Y(t) \tag{5.49}$$

where $(\hat{A}^0, \hat{C})$ is an observable pair, $Z(t)$ is the observed states.

It is proven that the observer Σ_0 is asymptotically stable (Hara, et al., 1976), i.e.,

$$\lim_{t\to\infty} \frac{d^j}{dt^j}(W(t) - X(t)) = 0 \qquad (j = 0,1,2,...) \tag{5.50}$$

and is independent on input $U(\cdot)$ and initial states $X(0)$, $Z(0)$. The parameters of the observer are given as follows

$$\hat{A}^i = 0 \qquad (i = 1,2,...,m)$$

$$\hat{B}^i = HA_{11}^i + A_{21}^i$$

$$\hat{A}^0 = A_{22}^0 + HA_{12}^0$$

$$\hat{B}^0 = HA_{11}^0 + A_{21}^0 - (A_{22}^0 + HA_{12}^0)H$$

$$\hat{J} = HB_1 + B_2$$

$$\hat{C} = \begin{pmatrix} 0 \\ I_{n-p} \end{pmatrix} \qquad \widehat{D} = \begin{pmatrix} I_p \\ -H \end{pmatrix}$$

5.3.2 State-disturbance composite observer

Consider an n-dimensional bilinear system Σ_p' with unmeasurable disturbances

$$\dot{X}(t) = A^0 X(t) + \sum_{i=1}^{m} A^i U_i(t) X(t) + BU(t) + FV(t) \tag{5.51}$$

$$Y(t) = CX(t) + DV(t) \tag{5.52}$$

where, $X \in R^n$, $U \in R^m$, $V \in R^l$, $Y \in R^p$ are state, control, disturbance, and output vectors, respectively; A^i $(i = 0, 1, 2, ..., m)$, B, C, D, F are corresponding coefficient matrices. The l-dimensional disturbance vector can be represented by the following disturbance state equations

$$\dot{X}_w(t) = A_w X_w(t) \tag{5.53}$$

$$V(t) = C_w X_w(t) \tag{5.54}$$

where, $X_w \in R^{n_w}$ is the disturbance state vector. If the disturbance vector consists of l_1 measurable disturbances and l_2 unmeasurable disturbances, i.e.,

$$V(t) = \begin{pmatrix} V_1(t) \\ V_2(t) \end{pmatrix} \tag{5.55}$$

where, $V_1(t) \in R^{l_1}$, $V_2(t) \in R^{l_2}$ are measurable and unmeasurable disturbance vectors respectively, and $(l_1 + l_2 = l)$. It follows that equation (5.55) can be rewritten as

$$V(t) = \begin{pmatrix} V_1(t) \\ V_2(t) \end{pmatrix} = \begin{pmatrix} C_{w_1} \\ C_{w_2} \end{pmatrix} X_w(t) \tag{5.56}$$

Then, the system Σ_p' can be represented by the following equations

$$\dot{\bar{X}}(t) = \bar{A}^0 \bar{X}(t) + \sum_{i=1}^{m} \bar{A}^i \bar{U}_i(t) \bar{X}(t) + \bar{B} \bar{U}(t) \tag{5.57}$$

$$\bar{Y}(t) = \bar{C} \bar{X}(t) \tag{5.58}$$

where

$$\bar{X} = \begin{pmatrix} X \\ X_w \end{pmatrix} \qquad \bar{U} = U \qquad \bar{Y} = \begin{pmatrix} Y \\ V_1 \end{pmatrix}$$

$$\bar{A}^0 = \begin{pmatrix} A^0 & FC_w \\ 0 & A_w \end{pmatrix} \qquad \bar{A}^i = \begin{pmatrix} A^i & 0 \\ 0 & 0 \end{pmatrix}$$

$$\bar{B} = \begin{pmatrix} B \\ 0 \end{pmatrix} \qquad \bar{C} = \begin{pmatrix} C & DC_w \\ 0 & C_{w_1} \end{pmatrix}$$

Let $\bar{n} = n + n_w$, $\bar{m} = m$, $\bar{p} = p + l_1$, and $rank(\bar{C}) = \bar{p}$, then there exists a nonsingular transformation matrix T $(\tilde{X} = T\bar{X})$ which can transform the system Σ_p' into Σ_p^{**} as follows

$$\dot{\tilde{X}}(t) = \begin{pmatrix} \tilde{A}_{11}^0 & \tilde{A}_{12}^0 \\ \tilde{A}_{21}^0 & \tilde{A}_{22}^0 \end{pmatrix} \tilde{X}(t) + \sum_{i=1}^{m} \begin{pmatrix} \tilde{A}_{11}^i & \tilde{A}_{12}^i \\ \tilde{A}_{21}^i & \tilde{A}_{22}^i \end{pmatrix} \bar{U}_i(t) \tilde{X}(t)$$

$$+ \begin{pmatrix} \tilde{B}_1 \\ \tilde{B}_2 \end{pmatrix} \bar{U}(t) \tag{5.59}$$

$$\bar{Y}(t) = (I_{\bar{p}} \quad 0)\tilde{X}(t) \qquad (5.60)$$

where, $\tilde{A}_{11}^i \in R^{\bar{p} \times \bar{p}}$, $\tilde{A}_{12}^i \in R^{\bar{p} \times (\bar{n}-\bar{p})}$, $\tilde{A}_{21}^i \in R^{(\bar{n}-\bar{p}) \times \bar{p}}$, $\tilde{A}_{22}^i \in R^{(\bar{n}-\bar{p}) \times (\bar{n}-\bar{p})}$, $(i = 0, 1, 2, ..., m)$, $\tilde{B}_1 \in R^{\bar{p} \times \bar{m}}$, $\tilde{B}_2 \in R^{(\bar{n}-\bar{p}) \times \bar{m}}$. Thus, the system Σ_p^{**} is similar to the system Σ_p^* and has no unknown inputs.

Theorem 3 *For the system Σ_p^{**}, if there exists an $(\bar{n} - \bar{p}) \times \bar{p}$ matrix $\tilde{H}$ satisfying the following two conditions:*
   (1) $\qquad \tilde{A}_{22}^i + \tilde{H}\tilde{A}_{12}^i = 0$
   (2) $\qquad$ *All eigenvalues of $(\tilde{A}_{22}^0 + \tilde{H}\tilde{A}_{12}^0)$ have negative real parts,*
then the state disturbance composite observer Σ_0^{**} can be constructed as follows:

$$\dot{\tilde{Z}}(t) = \hat{A}^0 \tilde{Z}(t) + \sum_{i=1}^{\bar{m}} \hat{A}^i \bar{U}_i(t) \tilde{Z}(t) + \hat{B}^0 \tilde{Y}(t)$$

$$+ \sum_{i=1}^{\bar{m}} \hat{B}^i \bar{U}_i(t) \tilde{Y}(t) + \hat{J}\bar{U}(t) \qquad (5.61)$$

$$\tilde{W}(t) = C\tilde{Z}(t) + D\tilde{Y}(t) \qquad (5.62)$$

Also, it can be proven that the composite observer Σ_0^{**} is asympototically stable, i.e.,

$$\lim_{t \to \infty} \frac{d^j}{dt^j}(\tilde{W}(t) - \bar{X}) = 0 \qquad (j = 0, 1, 2, ...) \qquad (5.63)$$

and is independent of both input $\bar{U}(\cdot)$ and initial states $\bar{X}(0)$ and $\bar{Z}(0)$. The parameter matrices of Σ_0^{**} are given as follows

$$\hat{A}^i = 0 \qquad (i = 1, 2, ..., \bar{m})$$

$$\hat{B}^i = \tilde{H}\tilde{A}_{11}^i + \tilde{A}_{21}^i \qquad (i = 1, 2, ..., \bar{m})$$

$$\hat{A}^0 = \tilde{A}_{22}^0 + \tilde{H}\tilde{A}_{12}^0$$

$$\hat{B}^0 = \tilde{H}\tilde{A}_{11}^0 + \tilde{A}_{21}^0 - (\tilde{A}_{22}^0 + \tilde{H}\tilde{A}_{12}^0)\tilde{H}$$

$$\hat{J} = \tilde{H}\tilde{B}_1 + \tilde{B}_2$$

$$\hat{C} = \begin{pmatrix} 0 \\ I_{\bar{n}-\bar{p}} \end{pmatrix} \qquad \hat{D} = \begin{pmatrix} I_{\bar{p}} \\ -\tilde{H} \end{pmatrix}$$

All of the eigenvalues of the minimal order composite observer can be assigned arbitrarily (Ying, et al., 1991)

The conditions in Theorem 2 can be further simplified. First, condition (1) can be rewritten as

$$\tilde{H}(\tilde{A}_{12}^1 \tilde{A}_{12}^2 \ ... \ \tilde{A}_{12}^{\bar{m}}) = (-\tilde{A}_{22}^1 - \tilde{A}_{22}^2 \ ... \ - \tilde{A}_{22}^{\bar{m}}) \qquad (5.64)$$

Let T_1 be a nonsingular transformation matrix that satisfies the condition

$$(\tilde{A}_{12}^1 \ \tilde{A}_{12}^2 \ ... \ \tilde{A}_{12}^{\tilde{m}})T = (G_1 \ \ 0) \tag{5.65}$$

where $rank(G_1) = rank(\tilde{A}_{12}^1 \ \tilde{A}_{12}^2 \ \ ... \ \ \tilde{A}_{12}^{\tilde{m}}) = r$. Right-multiplying (5.64) by T_1 yields

$$\tilde{H}(G_1 \ \ 0) = (Q_1 \ \ Q_2) \tag{5.66}$$

Since Q_1 has the same columns as G_1, condition (1) can be satisfied only when $Q_2 = 0$. As G_1 is of column full rank, we assume its first r rows to be nonsingular, i.e.,

$$G_1 = \begin{pmatrix} G_{11} \\ G_{12} \end{pmatrix} \tag{5.67}$$

where G_{11} is nonsingular. Similarly, $\tilde{H}$ can also be decomposed as

$$\tilde{H} = (\tilde{H}_1 \ \ \tilde{H}_2) \tag{5.68}$$

where $\tilde{H}_1$ is the first left r columns of $\tilde{H}$. If $Q_2 = 0$ holds, then $\tilde{H}$ can be obtained from equation (5.64)

$$\tilde{H} = ((Q_1 G_{11}^{-1} - \tilde{H}_2 G_{12} G_{11}^{-1}) \ \ \tilde{H}_2) \tag{5.69}$$

where $\tilde{H}_2$ is an arbitrary matrix. Substituting (5.69) into condition (2) in Theorem 3, we have

$$\begin{aligned} \tilde{A}_{22}^0 + \tilde{H}\tilde{A}_{12}^0 &= (\tilde{A}_{22}^0 + Q_1 G_{11}^{-1} \tilde{A}_{121}^0) \\ &+ \tilde{H}_2(\tilde{A}_{122}^0 - G_{21} G_{11}^{-1} \tilde{A}_{121}^0) \end{aligned} \tag{5.70}$$

where $\tilde{A}_{12}^0 = \begin{pmatrix} \tilde{A}_{121}^0 \\ \tilde{A}_{122}^0 \end{pmatrix}$ and $\tilde{A}_{121}^0$ has r rows.

Let

$$Q = \tilde{A}_{22}^0 + Q_1 G_{11}^{-1} \tilde{A}_{121}^0$$
$$G = \tilde{A}_{122}^0 - G_{21}^{-1} G_{11}^{-1} \tilde{A}_{121}^0$$

then, equation (5.70) becomes

$$\tilde{A}_{22}^0 + \tilde{H}\tilde{A}_{12}^0 = Q + \tilde{H}_2 G \tag{5.71}$$

Thus, we can obtain the following theorem according to the well known pole-assignment theory.

Theorem 4 *For the system Σ_p^{**}, if $Q_2 = 0$ holds, and (Q, G) is an observable pair, then the conditions in Theorem 2 are completely satisfied, and all of the eigenvalues of the minimal order composite observer can be assigned arbitrarily.*

5.3.3 Application example

The headbox control system discussed above can be described by the following bilinear model if the disturbance variables are included (Ying, et al., 1991)

$$\dot{X}(t) = A^0 X(t) + \sum_{i=1}^{2} A^i U_i X(t) + BU(t) + FW(t) \quad (5.72)$$

$$Y(t) = CX(t) \quad (5.73)$$

where

$$A^0 = \begin{pmatrix} -1.93 & 0 & 0 & 0 \\ 0.394 & -0.426 & 0 & 0 \\ 0 & 0 & -0.63 & 0 \\ 0.82 & -0.784 & 0.413 & -0.426 \end{pmatrix}$$

$$A^1 = A^2 = \begin{pmatrix} 0 & 0 & 0 & 0 \\ 0 & 0 & 0 & 0 \\ 0 & 0 & -0.327 & 0 \\ 0 & 0 & 0 & 0 \end{pmatrix} \quad B = \begin{pmatrix} 1.274 & 1.274 \\ 0 & 0 \\ 1.34 & -0.65 \\ 0 & 0 \end{pmatrix}$$

$$C = \begin{pmatrix} 0 & 1 & 0 & 0 \\ 0 & 0 & 1 & 0 \\ 0 & 0 & 0 & 1 \end{pmatrix} \quad F = \begin{pmatrix} 0 & 0 \\ 0 & 0 \\ 0.203 & 0.406 \\ 0 & 0 \end{pmatrix}$$

The state variables $(X = (H_1 \; H_2 \; N_1 \; N_2)^T)$ are the level of mixing box, the level of headbox, the consistency of stock in mixing box, and the consistency of stock in headbox, respectively . The disturbance variables $(W = (N_p \; N_w)^T)$ are the consistency of stock which comes from pulp section, as well as the consistency of white water. The control variables $(U = (G_p \; G_w)^T)$ are the flow rate of the stock, and the flow rate of white water. Among the state and disturbance variables, H_2, N_1, N_2 and N_p can be measured by on-line sensors, whereas H_1 and N_w need to be estimated by using an observer in order to design a state feedback and disturbance feedforward control system.

According to the disturbance state equations

$$\dot{X}_w(t) = A_w X_w(t) \quad (5.74)$$

$$W(t) = C_w X_w(t) \quad (5.75)$$

we have

$$A_w = 0 \quad C_w = \begin{pmatrix} 1 & 0 \\ 0 & 1 \end{pmatrix} \quad X_w(t) = \begin{pmatrix} N_p \\ N_w \end{pmatrix}$$

Let

$$\bar{X}(t) = \begin{pmatrix} X(t) \\ X_w(t) \end{pmatrix} \qquad \bar{U}(t) = U(t) \quad \bar{Y}(t) = \begin{pmatrix} Y(t) \\ W_1(t) \end{pmatrix}$$

the composite system becomes

$$\dot{\bar{X}}(t) = \bar{A}^0 \bar{X}(t) + \sum_{i=1}^{2} \bar{A}^i \bar{U}_i(t) \bar{X}(t) + \bar{B}\bar{U}(t) \tag{5.76}$$

$$\bar{Y}(t) = \bar{C}\bar{X}(t) \tag{5.77}$$

Take the transformation matrix as

$$T = \begin{pmatrix} 0 & 0 & 0 & 1 & 0 & 0 \\ 0 & 1 & 0 & 0 & 0 & 0 \\ 0 & 0 & 1 & 0 & 0 & 0 \\ 0 & 0 & 0 & 0 & 1 & 0 \\ 1 & 0 & 0 & 0 & 0 & 0 \\ 0 & 0 & 0 & 0 & 0 & 1 \end{pmatrix} \tag{5.78}$$

then $\tilde{X} = T\bar{X} = (N_2 \ H_2 \ N_1 \ N_p \ H_1 \ N_w)^T$. The system can be transformed into

$$\dot{\tilde{X}}(t) = \tilde{A}^0 \tilde{X}(t) + \sum_{i=1}^{2} \tilde{A}^i \bar{U}_i(t) \tilde{X}(t) + \tilde{B}\bar{U}(t) \tag{5.79}$$

$$\tilde{Y} = \tilde{C}\tilde{X}(t) \tag{5.80}$$

where

$$\tilde{A}^0 = \begin{pmatrix} \tilde{A}^0_{11} & \tilde{A}^0_{12} \\ \tilde{A}^0_{21} & \tilde{A}^0_{22} \end{pmatrix} = \begin{pmatrix} -0.426 & -0.784 & 0.413 & 0 & 0.818 & 0 \\ 0 & -0.426 & 0 & 0 & 0.394 & 0 \\ 0 & 0 & -0.63 & 0.203 & 0 & 0.406 \\ 0 & 0 & 0 & 0 & 0 & 0 \\ 0 & 0 & 0 & 0 & -1.93 & 0 \\ 0 & 0 & 0 & 0 & 0 & 0 \end{pmatrix}$$

$$\tilde{A}^i = \begin{pmatrix} \tilde{A}^i_{11} & \tilde{A}^i_{12} \\ \tilde{A}^i_{21} & \tilde{A}^i_{22} \end{pmatrix} = \begin{pmatrix} 0 & 0 & 0 & 0 & 0 & 0 \\ 0 & 0 & 0 & 0 & 0 & 0 \\ 0 & 0 & -0.327 & 0 & 0 & 0 \\ 0 & 0 & 0 & 0 & 0 & 0 \\ 0 & 0 & 0 & 0 & 0 & 0 \\ 0 & 0 & 0 & 0 & 0 & 0 \end{pmatrix} \qquad (i = 1, 2)$$

$$\tilde{B} = \begin{pmatrix} \tilde{B}_1 \\ \tilde{B}_2 \end{pmatrix} = \begin{pmatrix} 0 & 0 \\ 0 & 0 \\ 1.34 & -0.65 \\ 0 & 0 \\ 1.274 & 1.274 \\ 0 & 0 \end{pmatrix}$$

$$\tilde{C} = \begin{pmatrix} I_{\bar{p}} & 0 \end{pmatrix} = \begin{pmatrix} 1 & 0 & 0 & 0 & 0 & 0 \\ 0 & 1 & 0 & 0 & 0 & 0 \\ 0 & 0 & 1 & 0 & 0 & 0 \\ 0 & 0 & 0 & 1 & 0 & 0 \end{pmatrix}$$

$$\tilde{Y}(t) = (N_2 \ H_2 \ N_1 \ N_p)^T.$$

Obviously, this system satisfies condition (1) in Theorem 3. Condition (2) can also be satisfied by choosing an appropriate $\tilde{H}$. Here, we choose:

$$\tilde{A}_{22}^0 + \tilde{H}\tilde{A}_{12}^0 = \begin{pmatrix} -0.5 & 0 \\ 0 & -0.2 \end{pmatrix} \tag{5.81}$$

Then, $\tilde{H}$ can be obtained:

$$\tilde{H} = \begin{pmatrix} 1.744 & 0 & 0 & 0 \\ 0 & 0 & -0.49 & 0 \end{pmatrix} \tag{5.82}$$

Thus, we get a minimal order $(6 - 4 = 2)$ state-disturbance composite observer

$$\dot{\tilde{Z}}(t) = A^0 \tilde{Z}(t) + B^0 \tilde{Y}(t)$$

$$+ \sum_{i=1}^{2} B^i \bar{U}_i(t)\tilde{Y}(t) + J\bar{U}(t) \tag{5.83}$$

$$\tilde{W}(t) = C\tilde{Z}(t) + D\tilde{Y}(t) \tag{5.84}$$

where

$$A^0 = \begin{pmatrix} -0.5 & 0 \\ 0 & -0.2 \end{pmatrix}$$

$$B^1 = B^2 = \begin{pmatrix} 0 & 0 & 0 & 0 \\ 0 & 0 & 0.161 & 0 \end{pmatrix}$$

$$B^0 = \begin{pmatrix} 0.129 & -1.367 & 0.72 & 0 \\ 0 & 0 & 0.21 & -0.01 \end{pmatrix}$$

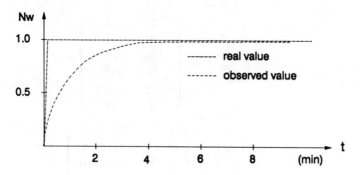

Figure 5.8: N_w with a unit step change and its observed value

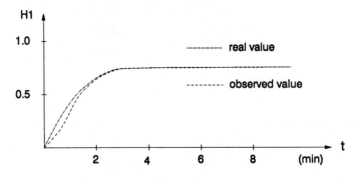

Figure 5.9: The values of H_1 with a unit step change in G_w

$$J = \begin{pmatrix} 1.274 & 1.274 \\ 0.659 & 0.315 \end{pmatrix}$$

$$C = \begin{pmatrix} 0 & 0 \\ 0 & 0 \\ 0 & 0 \\ 0 & 0 \\ 1 & 0 \\ 0 & 1 \end{pmatrix} \qquad D = \begin{pmatrix} 1 & 0 & 0 & 0 \\ 0 & 1 & 0 & 0 \\ 0 & 0 & 1 & 0 \\ 0 & 0 & 0 & 1 \\ -1.74 & 0 & 0 & 0 \\ 0 & 0 & 0.49 & 0 \end{pmatrix}$$

As we know, the output variables of the observer $\tilde{W}_5$ and $\tilde{W}_6$ are the estimated state variable H_1 and the disturbance variable N_w. The simulation results for this composite observer are shown in Figures 5.8 - 5.11.

As shown in Figures 5.8 and 5.9, the stability and tracking behavior of this composite observer are quite good. Both the estimated state and disturbance can reach their true values very quickly. Figures 5.10 and 5.11 indicate that when the disturbance N_w changes, the control quality of the headbox system

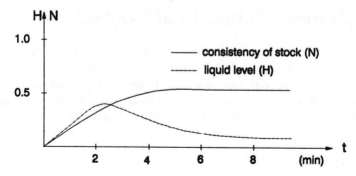

Figure 5.10: The system responses without feedforward compensation for N_w unit step change

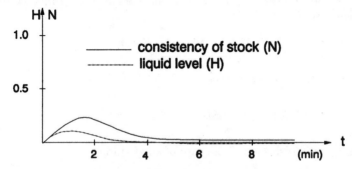

Figure 5.11: The system responses with feedforward compensation for N_w unit step change

with feedforward compensation by using a composite observer is significantly better than that without feedforward compensation. The application of this composite observer to an on-line headbox control system shows very satisfactory results.

5.4   Bilinear Suboptimal Control

In this section, the suboptimal control for bilinear systems is discussed by use of the extension of linear-quadratic optimal control index. The design method of this bilinear suboptimal control system is presented. Its application to the moisture control of a paper machine, as an example, is given. The simulation results show that this suboptimal control system functions very well.

5.4.1   Background

Since Mohler (1973) discussed bang-bang control and minimal time control for bilinear systems, many researchers have worked on BLS optimal control problem. Jacobson (1977) extended the linear-quadratic optimal controller to the following BLS

$$\dot{X} = AX + \sum_{i=1}^{m} B_i X u_i \tag{5.85}$$

and obtained the optimal control

$$u_i^*(X) = \frac{1}{2} X^T (S B_i + B_i S) X \qquad i = 1, 2, ..., m \tag{5.86}$$

Banks et al. (1986) discussed the case of single input

$$\dot{X} = AX + BXu, \qquad X(t_0) = X_0 \tag{5.87}$$

defining an objective function as

$$J(u) = \int_{t_0}^{t_f} u^2 dt + X^T(t_f) F X(t_f) \tag{5.88}$$

The Hamiltonian for the optimal control problem (5.88) subject to (5.87) is

$$H = u^2 + \lambda^T (AX + BXu) \tag{5.89}$$

The optimal control can be obtained as

$$u^* = -\frac{1}{2} \lambda^T B X \tag{5.90}$$

The problems discussed in the above literatures are optimal control of BLS in the special and simple cases. There are also some literatures (Derese and

Noldus, 1982; Ryan, 1984; Tzafestas, et al., 1984) reporting optimal control for a general BLS represented by

$$\dot{X} = AX + B_0 U + \sum_{i=1}^{m} B_i X u_i \tag{5.91}$$

However, the most results from these studies are quite difficult to solve the real industry problems.

In this section, the suboptimal control for BLS is discussed by using extension of the linear-quadratic optimal control. This kind of bilinear suboptimal control is easy to design and implement, and has good control performance.

5.4.2 Bilinear suboptimal control

Consider an n-dimensional bilinear system represented by

$$\dot{X} = AX + B_0 U + \sum_{i=1}^{m} B_i X U_i \qquad X(0) = X_0 \tag{5.92}$$

$$Y = CX \tag{5.93}$$

where, $X \in R^n$, $U \in R^m$, $Y \in R^p$ are state, input and output vectors respectively; A, B_0, B_i $(i = 1, 2, ..., m)$ and C are $n \times n$, $n \times m$, $n \times n$ and $p \times n$ matrices. Similar to linear system, the optimal control problem for bilinear system can be described as follows: for the system (5.92), if there exists a control law $U = U^*(x, t)$ that satisfies the following constraint

$$U_{min} \leq U^*(X, t) \leq U_{max} \tag{5.94}$$

and minimizes the performance index

$$J(U) = \int_0^\infty L(X, U, t)dt \tag{5.95}$$

then, $U^*(X, t)$ is an optimal control for the bilinear system (5.92).

From system (5.92), we know that the bilinear system is linear in state, linear in control, but not jointly. If $B_i = 0$ $(i=1,2,...,m)$, system (5.92) becomes a linear system. As we know, for linear system

$$\dot{X} = AX + BU \qquad X(0) = X_0 \tag{5.96}$$

if the system (A, B) is controllable, there is an optimal control

$$U^* = -R^{-1} B^T P X \tag{5.97}$$

which minimizes the LQ performance index

$$J(u) = \frac{1}{2}\int_0^\infty (X^T Q X + U^T R U)dt \qquad (5.98)$$

where Q is a positive definite matrix, R is a semi-positive definite matrix, and P matrix is the solution of the following Riccati equation

$$PA + A^T P - PBR^{-1}B^T P + Q = 0 \qquad (5.99)$$

For optimal control problem of bilinear system (5.92), we can use the same approach as the linear system. Let us discuss a single input case first. Consider a bilinear system

$$
\begin{aligned}
\dot{X} &= AX + B_0 u + B_1 X u & X(0) = X_0 \\
Y &= CX & (5.100)
\end{aligned}
$$

with LQ performance index

$$J(u) = \int_0^\infty (X^T Q X + u^T R u)dt \qquad (5.101)$$

where $Q \in R^{n \times n}$ and $R \in R^{1 \times 1}$ are positive definite and semi-positive definite matrices respectively. If we assume that the system (5.92) is controllable and its closed loop system is globally asymptotically stable under the action of optimal control u^*, then

$$\lim_{t \to \infty} X(t) = 0 \qquad (5.102)$$

$$X_0^T K X_0 + \int_0^\infty (X^T K \dot{X} + \dot{X}^T K X)dt = 0 \qquad (5.103)$$

where K is an arbitrary symmetric positive definite matrix (Derese and Noldus, 1982).

Combining equations (5.101) and (5.103), and eliminating the derivatives by using system equation (5.100), we obtain

$$
\begin{aligned}
J(u) &= X_0^T K X_0 \\
&+ \int_0^\infty [X^T Q X + u^T R u + X^T K (AX + B_0 u + B_1 X u) \\
&+ (AX + B_0 u + B_1 X u)^T K X]dt \qquad (5.104)
\end{aligned}
$$

Reorganizing the above equation yields

$$
\begin{aligned}
J(u) \;=\; & X_0^T K X_0 \\
& + \int_0^\infty [X^T(Q + KA + A^T K - K B_0 R^{-1} B_0^T K)X \\
& + (u + R^{-1}(B_0^T + X^T B_1^T)KX)^T R(u + R^{-1}(B_0^T + X^T B_1^T)KX) \\
& - (X^T K(B_1 X)R^{-1}(B_1 X)^T K X)]dt
\end{aligned}
\tag{5.105}
$$

If we select the appropriate K to satisfy

$$
KA + A^T K - K B_0 R^{-1} B_0^T K + Q = 0
\tag{5.106}
$$

equation (5.105) becomes

$$
\begin{aligned}
J(u) \;=\; & X_0^T K X_0 \\
& + \int_0^\infty [(u + R^{-1}(B_0^T + X^T B_1^T)KX)^T R(u + R^{-1}(B_0^T + X^T B_1^T)KX) \\
& - (X^T K(B_1 X)R^{-1}(B_1 X)^T K X)]dt
\end{aligned}
\tag{5.107}
$$

Since equation (5.106) is a standard Riccati equation, there must exist a symmetric positive definite solution if (A, B_0) is controllable.

From equation (5.107), if we modify the performance index as follows

$$
J(u) = \int_0^\infty (X^T Q X + u^T R u + X^T K(B_1 X)R^{-1}(B_1 X)^T K X)dt
\tag{5.108}
$$

then it can be obtained from equations (5.105)-(5.108) that

$$
\begin{aligned}
J(u) \;=\; & \int_0^\infty [(u + R^{-1}(B_0^T + X^T B_1^T)KX]^T R \\
& (u + R^{-1}(B_0^T + X^T B_1^T)KX)]dt
\end{aligned}
\tag{5.109}
$$

From equation (5.109), the suboptimal control can be obtained

$$
u^*(x) = -R^{-1}(B_0^T + X^T B_1^T)KX
\tag{5.110}
$$

or rewritten as

$$
u^*(x) = u_L^*(x) + u_B^*(x) = -R^{-1}B_0^T K X - R^{-1}X^T B_1^T K X
\tag{5.111}
$$

Obviously, the $u_L^*(x)$ is the linear-quadraic optimal control, and $u_B^*(x)$ is the modification control for nonlinear (bilinear) part of the systems.

The result obtained above can be applied to multi-input cases. Let us consider a bilinear system with m inputs as follows

$$\dot{X}(t) = AX(t) + B_0 U + \sum_{i=1}^{m} B_i X U_i \qquad (5.112)$$

where, $X \in R^n$, $U \in R^m$, $A \in R^{n \times n}$, $B_0 \in R^{n \times m}$, and $B_i \in R^{n \times n}$.

Defining
$$B_x = (B_1 X \ B_2 X \ ... \ B_m X)$$

equation (5.112) can be rewritten as

$$\dot{X} = AX + (B_0 + B_x)U \qquad (5.113)$$

From the above discussion, we can apply a suboptimal control for system (5.113)

$$U^* = -R^{-1}(B_0^T + B_x^T)KX \qquad (i = 1, 2, ..., m) \qquad (5.114)$$

where K satisfies the following Riccati equation

$$KA + A^T K - K B_0 R^{-1} B_0^T K + Q = 0 \qquad (5.115)$$

where Q is an arbitrary symmetric positive definite matrix and R is a semi-positive $m \times m$ matrix.

5.4.3   Application

The dryer section of paper machine can be described by the following bilinear system (Ying, et al., 1992)

$$\dot{X} = AX + B_0 u + B_1 X u \qquad (5.116)$$

$$Y = CX \qquad (5.117)$$

where
$$A = \begin{pmatrix} -0.046 & 0 \\ -0.7632 & -3.197 \end{pmatrix} \qquad B_0 = \begin{pmatrix} 0.986 \\ 0 \end{pmatrix}$$

$$B_1 = \begin{pmatrix} -0.27 & 0 \\ 0 & 0 \end{pmatrix} \qquad C = \begin{pmatrix} 1 & 0 \\ 0 & 1 \end{pmatrix}$$

The bilinear suboptimal control system can be designed as shown in Figure 5.12. The corresponding suboptimal control law is

$$U^* = -R^{-1}(B_0^T + X^T B_1^T)KX \qquad (5.118)$$

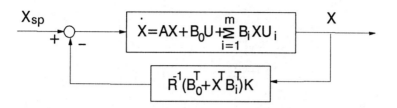

Figure 5.12: Bilinear suboptimal system

where K is a symmetric positive definite matrix and the solution of the following Riccati equation

$$KA + A^T K - KB_0 R^{-1} B_0^T K + Q = 0 \tag{5.119}$$

with

$$Q = \begin{pmatrix} 1 & 0 \\ 0 & 1 \end{pmatrix} \qquad R = 1$$

By solving the Riccati equation (5.119), we obtain

$$K = \begin{pmatrix} k_{11} & k_{12} \\ k_{21} & k_{22} \end{pmatrix} = \begin{pmatrix} 0.9892 & -0.0284 \\ -0.0284 & 0.1563 \end{pmatrix} \tag{5.120}$$

The suboptimal control is

$$U^*(x) = -0.975x_1 + 0.267x_1^2 + 0.028x_2 - 7.67 \times 10^{-3} x_1 x_2 \tag{5.121}$$

Figure 5.13 is the system block diagram. The simulation result of system performance is given in Figure 5.14, which shows that both the stability and performance of the system are quite satisfactory.

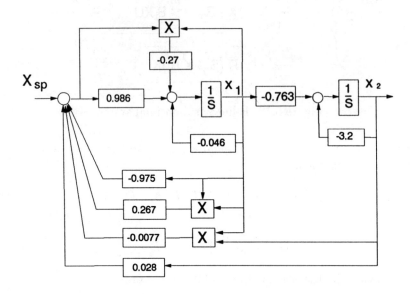

Figure 5.13: Bilinear suboptimal system for moisture control

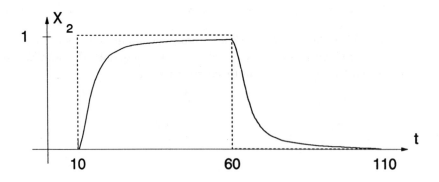

Figure 5.14: System response to unit step change in setpoint

5.5 Conclusions

In this Chapter, the state feedback decoupling control system for bilinear multivariable processes was discussed. The necessary condition for the existence and the design method of bilinear multivariable decoupling controller were proposed. The resulting control system was not only decoupled but also arbitrarily pole-assigned.

A state-disturbance composite observer for a class of bilinear systems and its design procedure were discussed. The composite observer for the headbox control system of a paper machine was designed. The simulation results and operation observations showed that: (a) the composite observer has very good stability and tracking behaviors; (b) both states and disturbances can be estimated at the same time. The feedforward compensation using estimated disturbances can improve the quality of control system significantly.

We had also studied the optimal control problem for bilinear systems. A bilinear-quadratic suboptimal algorithm was proposed. This algorithm was easy to implement and with satisfactory performance.

The proposed bilinear observer and control algorithms are of great potential to be applied to pulp and paper as well as other process industries.

5.6   References

Banks S.P. and Yew M.K. On the optimal control of bilinear systems and its relation to Lie algebras. Int. J. Control 1986; 43(:891-900

Bruni C., Dipillo G. and Koch G. Bilinear systems: an appealing class of "nearly linear" systems in theory and applications. IEEE Tans. Autom. Contr. 1974; AC-19:334-348

Derese I. and Noldus E. Optimization of bilinear control systems. Int. J. Systems Sci. 1982; 13:237-246

Derese I., Stevens P. and Noldus E. Observers for bilinear systems with bounded input. Int. J. Systems Science 1979; 10(6):649-668

Freund E. The structure of decoupled nonlinear system. Int. J. Control 1975; 21:443-450

Freund E. Decoupling and pole assignment in nonlinear systems. Electronics Letters 1973; 9:373-375

Funahashi Y. A class of state observers for bilinear systems. Int. J. Systems Sci. 1978; 9:1199-1205

Hara S. and Furuta K. Minimal order state observers for bilinear systems. Int. J. Control 1976; 24:705-718

Hsu C.S. and Karanam V.R. Observer design of bilinear systems. J. of Dyn. Syst. Meas. and Contr. 1983; 105:206-208

Isidori A. Nonlinear Control Systems: An Introduction. Springer-Verlag, New York, 1985

Isidor A., Krener A.J., Gori Giorgi C. and Monaco S. Nonlinear decoupling via feedback: A differential geometric approach. IEEE Trans. Autom. Contr. 1981; AC-26:331-345

Jacobson D.H. Extension of linear-quadratic control optimization and matrix theory. Academic Press, San Francisco, 1977

Li C.W. and Feng Y.K. Decoupling theory of general multivariable analytic non-linear systems. Int. J. Control 1987; 45:1147-1160

Maghsoodi Y. Design and computation of near-optimal stable observers for bilinear systems. IEEE Proceedings 1989; 136:127-132

Mohler R.R. Bilinear Control Processes. Academic Press New York and London, 1973

Mohler R.R. and Kolodziej W.J. An overview of bilinear system theory and applications. IEEE Trans. Syst. Man and Cyb. 1980; 10:683-689

Mohler R.R. Nonlinear system – Applications to bilinear control. Prentice Hall, Englewood Cliffs, 1991

Nazar S. and Rekasius Z.V. Decoupling of a class of nonlinear system. IEEE Trans. Autom. Contr. 1971; AC-16:257-260

Nijmeijer H. and Vander Schaft A.J. Nonlinear dynamical control systems. Springer-Verlag, New York, 1990

Rao R., Ying Y. and Corbin J. Intelligent engineering approach to pulp and paper process control. Proceeding of CPPA, Montreal, Canada, 1991, pp A195-A199

Ryan E.P. Optimal feedback control of bilinear systems. J. of Optimization Theory and Applications 1984; 44:333-362

Sinha P.K. State feedback decoupling of nonlinear system. IEEE Trans. Autom. Contr. 1977; AC-22:487-489

Sinha P.K. Multivariable control – An introduction. Marcel Dekker Inc., New York and Basel, 1984

Tzafestas S.G., Angnostou K.E. and Pimenides T.G. Stabilizing optimal control of bilinear system with a generalized cost. Optimal Control Appl. and Methods 1984; 5:111-117

Ying Y., Rao M. and Sun Y. State-disturbance composite observer for bilinear systems. Proc. of American Control Conference, San Diego, CA, 1990, pp 1917-20

Ying Y., Rao M. and Sun Y. Bilinear control strategy for paper-making process. Chemical Engineering Communications 1992; VIII:13-28

Ying Y., Rao M. and Sun Y. Bilinear state-disturbance composite observer and its application. Int. J. Systems Sci. 1991; 22:2489-2498

Ying Y., Rao M. and Sun Y. A new design method for bilinear suboptimal systems. Proc of American Control Conference, Boston, Massachusetts, 1991, pp 1820-1822

CHAPTER 6
FAULT-TOLERANT CONTROL

The increasing complexity of modern engineering production systems and requirements for high quality products have created a need for control systems with fault-tolerance. This chapter presents two algorithms for fault-tolerance analysis and fault-tolerant controller design. In Section 6.2, a model-based fault detection and fault-tolerant control technique for a pressurized headbox is presented. The sensor failures are detected and then located. The controller and state estimator are automatically reorganized subsequently to the occurrence of the failures to ensure the stability and acceptable performance of the closed loop system. In Section 6.3, a linear quadratic optimal system with the highest fault-tolerance is developed for the drying section. Quantitative relationship between fault-tolerance and controller designing parameters are established. An iterative algorithm is proposed to design control system with the highest fault-tolerance.

6.1   Introduction

Modern pulp and paper processes rely on evermore sophisticated control functions to meet the increasing performance requirements. The control systems are highly vulnerable to incidents such as failures of components (actuators, sensors and process components).

To increase the reliability in the presence of such incidents, the control system will have to provide fault-tolerance function. Fault-tolerant control system can adapt to significant changes of environments. When all components are in normal conditions, the system has the desired performance. If some key components (e.g., sensors and actuators) fail, the fault-tolerant control system still holds acceptable performance (Siljak, 1980). Fault-tolerant control is a

157

challenge to the classical control techniques. For the past two decades, fault-
tolerant control systems have attracted significant research efforts (Eterno, et
al., 1985; Shimenura and Fujita, 1985; Mariton and Bertrand, 1987). The
advances of computer system and parallel computing technology have made it
possible to realize the fault-tolerant control techniques in industrial plants.

One way to achieve fault-tolerance is to use redundant actuators and sensors. The redundancy management system performs detection and isolation
of the failures, reconfigurates the remaining components and control and filtering algorithms (Looze, et al., 1984; Xia and Rao, 1992). An alternative
is to design systems with fault tolerant structure and parameters. Petersen
(1987) designed a single controller which simultaneously stabilized a collection
of linear systems. Vidyasagar and Viswandaham (1985) proposed a method
to stabilize a single plant by a collection of failure-prone linear controllers.
Fault-tolerance analysis and fault-tolerant controller design have become a
major research topic in control theory. There have been the extensive simulation and application studies on fault tolerant control for flight control (Deckert
et al., 1977), but very few have been reported in pulp and paper industry.

This chapter presents two algorithms of fault-tolerance analysis and fault-
tolerant controller design for paper machines. A model-based fault detection
and fault-tolerant control technique for the pressurized headbox of a paper machine is presented in Section 6.2. A bank of Kalman filters is constructed with
respect to all the possible sensor failure modes. The possibility of each failure
mode hypothesis is calculated using measurement innovation processes. The
sensor failures are detected and then located based on the calculated possibil-
ities of the hypotheses. The controller and state estimator are automatically
reorganized subsequently to the occurrence of failures to ensure the stability
and acceptable performance of the closed-loop system. The issues of system
hardware redundancy and computational burden as well as implemental complexity are taken into account in the system design. Simulation results have
shown satisfactory performance of the headbox control system after applying
the presented technique. A linear quadratic optimal system with high fault-
tolerance is developed in Section 6.3, for drying section of a paper machine.
The fault-tolerance of a system is defined as the allowable bound of sensor
and actuator gain degradations to maintain stability. Quantitative results for
fault-tolerance are obtained using Lyapunov matrix equation solutions. A relation is established between the fault-tolerance and the prespecified stability
degree σ as well as the weighting matrix Q in the cost function. An iterative
algorithm is developed to design the system with the highest fault-tolerance.

6.2 Fault-Tolerant Control of Headboxes

Paper machine headbox is one of the critical units of a pulp and paper process. A good control of headbox is extremely important to improve pulp and paper quality and to increase economic profit. Many sophisticated control algorithms have been successfully developed for paper machines headboxes, such as adaptive control (Borisson, 1979; D'Huster, et al., 1983), nonlinear control (Gunn and Sinha, 1983), multivariable optimal control with a reference model (Lebeau, et al., 1980), and optimal decoupling control (Xia, et al., 1988).

Those algorithms, however, are developed under the assumption that all the system components are of high reliability and will not fail. Once a failure occurs, the system performance will degrade and the system may even diverge. However, some components, especially the measurement instruments (sensors) used in the headboxes of paper machines are often poor in reliability. The control systems for headbox are required to deal with the problem of component failures.

This section investigates the fault detection and fault-tolerant control of headboxes. A control technique which is tolerant of sensor failures is proposed. It has been applied to a practical pressurized headbox. The technique has the moderate computational burden. It is easy to be implemented in other industrial processes.

6.2.1 Process specification

The pressurized headbox is a volume chamber. Its main function is to distribute the water-fiber suspension onto the fourdrinier wire as evenly and steadily as possible. The lower part of the headbox is filled with pulp stock and above stock is air. Inside the headbox is a rectifier which provides the even distribution of pulp fiber onto the fourdrinier wire.

Figure 2.1 (in Chapter 2) is a schematic diagram of the pressurized headbox. The pulp stock is pumped into the headbox by a fan pump and flows onto the wire through a slice lip. The flow rate of pulp stock into the headbox is controlled by fan pump speed. The air is circulated in the air chamber of the headbox with a vacuum pump. The air pressure in the headbox is controlled by a vacuum valve located in the suction side of the vacuum pump. A vacuum breaker, a pressure regulator and a manual valve are also installed in order to protect the vacuum pump, to limit the maximal value of pressure in the headbox, and adjust the opening of the vacuum control valve under normal operating condition.

The two most important control objectives are:

1. to maintain the stock level in the headbox: the optimum operating condition for the rectifier roll is the stock level just above the top of the rectifier. Therefore, the stock level must be hold constant;

2. to maintain the rush/drag ratio: the rush/drag ratio refers to the ratio of the jet velocity of slurry onto the wire to the speed of the wire. In order to improve the physical properties of pulp and paper product, the ratio should be maintained at a specific value. For the real industrial headbox investigated in this book, the desired rush/drag ratio is 1.02.

The jet velocity of slurry is proportional to the square root of the total head. To maintain a constant rush/drag ratio, the total head must be adjusted with the variation of wire speed (machine rate). Hence, the directly controlled variables are total head and stock level.

Based on mechanism analysis, the transfer function relating stock level to fan pump speed and that relating pressure to vacuum valve position are both of first-order if there were no interactions between the pressure and the level. Stock level variation has no steady state effect on air pressure, and neither does air pressure variation on total head. These facts can be explained by mass balance and gas axiom. The increase of stock level causes air pressure to increase and subsequently causes the outflow rate of air to increase until the air pressure regains its originative value. Actually, the steady state value of air pressure can be only affected by vacuum valve opening. Total head is the summation of air pressure and stock level. When air pressure increases, total head will increase simultaneously. However, the increase of total head will cause the outflow rate of stock to increase. As a result, liquid level decreases and total head regains its original value. Thus, the steady state value of total head can be only affected by the fan pump speed.

According to the mechanism analysis above, the block diagram of the pressurized headbox can be depicted as Figure 2.2 (in Chapter 2), with L, stock level; P, air pressure; U_1 fan pump speed; U_2 vacuum valve opening (Xia, et al., 1991), and

$$T_1 s L(s) + L(s) = K_1 U_1(s) - P(s) \qquad (6.1)$$

$$T_2 s P(s) + P(s) = K_2 U_2(s) + K_3 T_2 s L(s) \qquad (6.2)$$

The real industrial pressurized headbox investigated here is of sizes 235 inches wide, 48 inches deep and 57 inches high. A number of bump tests have been carried out in the headbox. Using the data obtained from bump tests, it is calculated that $K_1 = 1.75$, $K_2 = 1.54$, $K_3 = 13.12$, $T_1 = 0.2308$ and $T_2 = 0.1710$. The continuous state equation is thus

$$\dot{L}(t) = -4.33 L(t) - 4.33 P(t) + 7.58 U_1(t) \qquad (6.3)$$

$$\dot{P}(t) = -56.8L(t) - 62.7P(t) + 99.5U_1(t) + 9.01U_2(t) \qquad (6.4)$$

By integrating the differential equations (6.3) and (6.4) over each sampling interval T=0.1min, the following discrete equations of the headbox are obtained

$$L(k+1) = 0.905L(k) - 0.0629P(k) + 0.166U_1(k) - 0.049U_2(k) \qquad (6.5)$$

$$P(k+1) = 0.0586L(k) - 0.825P(k) + 1.44U_1(k) + 0.180U_2(k) \qquad (6.6)$$

Noting that the total head H is the summation of L and P, we have

$$H(k+1) = -0.0042H(k) + 0.148L(k) + 1.61U_1(k) + 0.138U_2(k) \qquad (6.7)$$

$$L(k+1) = -0.0629H(k) + 0.9682L(k) + 0.166U_1(k) - 0.049U_2(k) \qquad (6.8)$$

we rewrite the above equations into state space representation by defining the state vector $x(k) = (H(k)\ L(k))^T$, the output vector $y_r(k) = x(k)$ and the control vector $u(k) = (U_1(k)\ U_2(k))^T$

$$x(k) = \begin{pmatrix} -0.0042 & 0.148 \\ -0.0629 & 0.968 \end{pmatrix} x(k) + \begin{pmatrix} 1.61 & 0.131 \\ 0.166 & -0.049 \end{pmatrix} u(k) + w(k) \qquad (6.9)$$

$$y_r(k) = x(k) \qquad (6.10)$$

where $w(k)$ is process noise. The stock level, air pressure and total head in the headbox are measured on-line. Denoting $y(k) = (H(k)\ L(k)\ P(k))^T$, the measurement equation is

$$y(k) = \begin{pmatrix} 1.0 & 0.0 \\ 0.0 & 1.0 \\ 1.0 & -1.0 \end{pmatrix} x(k) + v(k) \qquad (6.11)$$

where $v(k)$ is measurement noise.

Since there are serious interactions between the control of the two output variables, H and L, a decoupling control law is required

$$u(k) = Fx(k) + Gv_m(k) \qquad (6.12)$$

where $v_m(k)$ represents set-points.

The problem of the headbox control is that the measurement instruments including liquid level sensor, air pressure sensor and total head sensor are not reliable enough and may fail during operation. When any of the sensors fails, the measured data will no longer represent the true operating conditions of the headbox. The control systems without considering the possible failures will lead to a poor performance. To improve the reliability and dynamic performance of the headbox control system, failure detection and control strategy modification are necessary. In the following sections, we will deal with the problems.

6.2.2   Sensor fault detection and state estimation

An essential prerequisite for the development of fault-tolerant control is an early and accurate process fault detection. A great number of failure detection techniques have been developed in the past two decades, such as failure-sensitive filter, generalized likelihood ratio test and multiple Kalman filter with a standard multiple hypothesis testing (Willsky, 1976). The design of failure detection systems for industrial processes involves the consideration of the following issues:

1. Industrial processes may involve slow change and abrupt change failures. The failure detection schemes must be applicable to both kinds of the failures.

2. Industrial processes usually do not possess many back-up components. The requirement for costly hardware redundancy should not be too high.

3. Industrial processes usually apply computers with relatively small storage and low computing speed. The scheme should have reasonable computational complexity.

4. Industrial process models are usually approximate to real process dynamics. The failure detection schemes are required to be robust to model errors.

5. The system should be easy to reorganize in order to retain integrity in the presence of failures.

According to the requirements above, a fault detection technique is developed.

Consider the linear discrete stochastic system with system dynamics

$$x(k+1) = A(k)x(k) + B(k)u(k) + w(k) \qquad (6.13)$$

$$y(k) = C(k)x(k) + v(k) \qquad (6.14)$$

where $x \in R^n$, $y \in R^m$ and $u \in R^r$, w and v are zero-mean, independent, white Gaussian sequences with the covariances defined by

$$E[w(k)w^T(j)] = Q(k)\delta_{kj} \qquad E[v(k)v^T(j)] = R(k)\delta_{kj} \qquad (6.15)$$

where δ_{kj} is the Kronecker delta. $Q(k)$ and $R(k)$ are positive symmetric matrices. Suppose M ($M \leq m$) out of m sensors are of poor reliability and tend to fail. Without the loss of generality, let us assume that the unreliable sensors correspond to the 1st, 2nd, $\cdots$, Mth elements of the measurement vector y, respectively. We first construct a number of measurement equations under different failure hypotheses, and then design Kalman filters based on the system dynamics and the measurement equations. The sensor failure can be detected by evaluating the performance of the Kalman filters.

Scheme 1: single-failure case

In this scheme, our attention is restricted to single-failure case, i.e., it is assumed that there will not be two or more sensors fail at the same time.

For a system with M unreliable sensors, we have $M + 1$ hypotheses to consider: the normal condition and when the ith $(i = 1, \cdots, M)$ sensor has failed.

Under the normal operating condition

$$H_0: \quad y(k) = C_0(k)x(k) + v(k) + d_0(k) \tag{6.16}$$

where $C_0(k) = C(k)$ and $d_0(k) = 0$.

When the ith sensor fails

$$H_i: \quad y(k) = C_i(k)x(k) + v(k) + d_i(k) \quad i = 1, 2, \cdots, M \tag{6.17}$$

where $d_i(k) = (0 \cdots 0 \ m_i(k) \ 0 \cdots 0)^T$. $C_i(k)$ is $C(k)$ matrix with the row corresponding to the ith sensor replaced by zero. $m_i(k)$ is the unknown mathematical expectation of the output of the ith sensor in failure. We shall solve the problem as if $m_i(k)$ were known.

There is usually analytical redundancy among dissimilar measurement instruments in industrial control systems. That is to say, not only the system (6.13) and (6.14) is observable, but it holds observable when one of the unreliable sensors has failed. Thus, $(C_i(k), A(k))$ is observable for all $i = 0, 1, \cdots, M$.

The most probable hypothesis will be selected based on a finite set of measurements, $Y(k) = \{y(1), y(2), \cdots, y(k)\}$. In order to construct $M + 1$ Kalman filter, we denote this filter by KF_0, KF_1, $\cdots$, KF_M, respectively, based on the measurement equations (6.16) and (6.17)

$$\hat{x}_i(k + 1|k) = A(k)\hat{x}_i(k) + B(k)u(k) \tag{6.18}$$

$$\hat{x}_i(k) = \hat{x}_i(k|k - 1) + K_i(k)\gamma_i(k) \tag{6.19}$$

$$\gamma_i(k) = y(k) - C_i(k)\hat{x}_i(k|k - 1) - d_i(k) \quad i = 0, 1, \cdots, M \tag{6.20}$$

where $\hat{x}_i$ $(i = 0, 1, \cdots, M)$ is the state estimation under the hypothesis H_i. The filter gain $K_i(k)$ is calculated from the equations

$$P_i(k+1|k) = A(k)P_i(k)A^T(k) + Q(k) \tag{6.21}$$

$$K_i(k) = P_i(k|k-1)C_i^T(k)[C_i(k)P_i(k|k-1)C_i^T(k) + R(k)]^{-1} \tag{6.22}$$

$$P_i(k) = P_i(k|k-1) - K_i(k)C_i^T(k)P_i(k|k-1) \tag{6.23}$$

According to Bayes' rule, $p_i(k)$, the probability that H_i is true, can be determined by the measurements:

$$p_i(k+1) = \frac{F_i(\gamma_i(k+1))p_i(k)}{\sum_{j=0}^M F_j(\gamma_j(k+1))p_j(k)} \qquad i = 0, 1, \cdots, M \tag{6.24}$$

where $F_i(\gamma_i(k+1))$ is the probability density of $\gamma_i(k+1)$ under the hypothesis H_i. Obviously, under the condition that the hypothesis H_i is true, $\gamma_i(k)$ is the zero-mean, Gaussian innovation process with the covariance

$$\begin{aligned} V_i(k) &= E(\gamma_i(k)\gamma_i^T(k)) \\ &= C_i(k)P_i(k|k-1)C_i^T(k) + R(k) \end{aligned} \tag{6.25}$$

$$F_i(\gamma_i(k+1)) = \frac{exp[-\frac{1}{2}\gamma_i^T(k+1)V_i^{-1}(k)\gamma_i(k+1)]}{(2\pi)^{\frac{m}{2}}[det(V_i(k))]^{\frac{1}{2}}} \tag{6.26}$$

where $det(\cdot)$ denotes the determinant of a matrix.

In equation (6.20), the true value of $m_i(k)$ in $d_i(k)$ is not available. It can be estimated from the sample mean of the output of the ith sensor. Note that because of the special structure of the matrix $C_i(k)$, the unknown mean $m_i(k)$ does not involve in the filter equation. That is, the estimate of the hypothesis conditioned filter will be correct.

The probability that H_i is true implies the performance of the filter KF_i. If no failure occurs, the innovation processes of all the $M+1$ Kalman filters are zero-mean Gaussian processes with bounded covariances. Thus, $p_i(k)$ for all $i = 0, 1, \cdots, M$ have a lower bound, i.e., $p_i(k) \geq \varepsilon$. Here $0 < \varepsilon < 1$ is a suitably chosen constant. When the ith sensor fails, H_i is the only true hypothesis. The innovation processes of all Kalman filters except KF_i will no longer be zero-mean and their covariances will increase because of the wrong measurement data. The performance of those filters will degrade dramatically. As a consequence, $p_j(k)$ decreases. Thus $p_i(k) \gg p_j(k)$ and $p_j(k) < \varepsilon$ for all $j \neq i$.

Based on the discussion above, the algorithm for fault detection and state estimation can be expressed as follows:

1. For some $i \neq 0$, if the following conditions hold

$$p_i(k) \gg p_j(k) \qquad \forall j \neq i \tag{6.27}$$

and

$$Max\{p_0(k), \cdots, p_{i-1}(k), p_{i+1}(k), \cdots, p_M(k)\} < \varepsilon \tag{6.28}$$

then the ith sensor has failed. All the Kalman filters except KF_i are not more effective. The final state estimate is the estimate of the filter KF_i

$$\hat{x}(k) = \hat{x}_i(k) \tag{6.29}$$

2. If there is no i ($i \neq 0$) that satisfies the conditions (6.27) and (6.28), then no sensor has failed. The system is in the normal condition. The final state estimate is a weighted sum of the state estimate of all Kalman filters. Since $p_i(k)$ represents the relative accuracy of the state estimate of each Kalman filter, and noting that

$$\sum_{i=0}^{M} p_i(k) = 1 \tag{6.30}$$

$p_i(k)$ can be served as the weighting factor in the summation. The final state estimate is given by

$$\hat{x}(k) = \sum_{i=1}^{M} p_i(k)\hat{x}_i(k) \tag{6.31}$$

The value of the lower bound ε outer is chosen according to the practical situation of the systems.

Scheme 2: multiple-failure case

When two or more sensors have failed, none of the Kalman filters in Scheme 1 are designed under true hypothesis. All the Kalman filters are driven by wrong measurements. Therefore, Scheme 1 cannot be applied to multiple-failure case. In order to detect multiple failures, we design a bank of filters such that each sensor failure only affects one filter.

$M + 1$ hypotheses are constructed as follows:

The normal condition

$$H_0: \quad y(k) = \bar{C}_0(k)x(k) + v(k) + d_0(k) \tag{6.32}$$

where $\bar{C}_0(k) = C(k)$ and $d_0(k) = 0$.

When all the unreliable sensors except the ith ($i = 1, \cdots, M$) one have failed

$$H_i: \quad y(k) = \bar{C}_i(k)x(k) + v(k) + d_i(k) \quad i = 1, 2, \cdots, M \tag{6.33}$$

where $d_i(k) = (m_1(k) \cdots m_{i-1}(k) \; 0 \; m_{i+1}(k) \cdots m_M(k) \; 0 \cdots 0)^T$. $\bar{C}_i(k)$ is $C(k)$ matrix with the rows corresponding to all the unreliable sensors except the ith one replaced by zero. $m_j(k)$ is the unknown mathematical expectation of the output of the jth sensor in failure. If $(\bar{C}_i(k), \; A(k))$ is observable for all $i = 0, 1, \cdots, M$, then just as in Scheme 1, $M+1$ Kalman filters can be designed based on the system dynamics (6.13) and the measurement equations (6.32) and (6.33). Because of the special structure of the matrix $\bar{C}_i(k)$, the failure of the ith sensor only affect the filters KF_0 and KF_i. Multiple failures can be detected by evaluating the performance of the Kalman filters.

The filter equations and the calculation of $p_i(k)$ are similar to those in Scheme 1. Defining $J = \{j_1, j_2, \cdots, j_s\} \subset \{1, 2, \cdots, M\}$, the algorithm for fault detection and state estimation is expressed as follows

1. If

$$p_0(k) < \varepsilon \tag{6.34}$$

and

$$p_i(k) < \varepsilon \qquad \forall i \in J \tag{6.35}$$

then the j_1th, j_2th $\cdots$, j_sth sensors have failed. The final state estimate is

$$\hat{x}(k) = \sum_{i \in J}^{M} p_i(k)\hat{x}_i(k)/(1 - p_0 - \sum_{j \in J} p_j(k)) \tag{6.36}$$

2. If there is no subset $J \subset \{1, 2, \cdots, M\}$ that satisfies (6.34) and (6.35), then no sensor has failed. The system is under normal condition. The final state estimation is

$$\hat{x}(k) = \sum_{i=0}^{M} p_i(k)\hat{x}_i(k) \tag{6.37}$$

Obviously, Scheme 2 requires more sensor redundancy than Scheme 1. Whether choosing Scheme 1 or Scheme 2 depends on the practical situation.

Scheme 3: reduced-order technique

As pointed out in Scheme 1, the output of the ith sensor does not participate in the state estimation of the filter KF_i because of the special structure of the matrix $C_i(k)$. For the sake of convenience in industrial applications, it is preferable to design M reduced-order Kalman filters, denoted by $KF_1, \; KF_2, \; \cdots, \; KF_M$ respectively, based on the system equation (6.13) and measurement equation (6.38)

$$y_i(k) = \hat{C}_i(k)x(k) + v_i(k) \qquad i = 1, 2, \cdots, M \tag{6.38}$$

where $y_i(k)$ is $y(k)$ vector with the element corresponding the ith unreliable sensor deleted, $\hat{C}_i(k)$ and $v_i(k)$ are the corresponding measurement matrix and measurement noise vector. The estimate error sequences γ_i are

$$\gamma_i(k) = y_i(k) - \hat{C}_i \hat{x}_i(k) \qquad i = 1, 2, \cdots, M \qquad (6.39)$$

It is clear from equations (6.38) and (6.39) that the dimensions of the measurement vectors have been reduced to $m-1$. The estimation of the sensors in failure in mathematical expectation is also avoided. Thus the computational burden is reduced.

The multiple-filter failure diagnosis technique intends to calculate the prediction error $\gamma_i(k+1)$ to measure how well each Kalman filter works. $p_i(k)$ is a simple measure of how well each Kalman filters works relative to each other and of how well we would expect them to work. Since the performance of the Kalman filters is assumed to be dependent on the performance of the sensors, $p_i(k)$ is also a measure of how well each group of sensors work. In the normal condition, the filter KF_0 should give the best result since it utilizes most of measurement information. When any of the unreliable sensors fails, the performance of KF_0 will be degraded. Only one Kalman filter KF_i $(i \neq 0)$ will still work well. Thus we can use the prediction errors of the filters KF_1, KF_2, $\cdots$, KF_M to detect failures and use the estimate of the filter KF_0 as the final state estimate in the normal condition.

A new failure detection scheme is given which is a little different from Scheme 1.

We first calculate $p_i(k)$

$$p_i(k+1) = \frac{F_i(\gamma_i(k+1))p_i(k)}{\sum_{j=1}^{M} F_j(\gamma_j(k+1))p_j(k)} \qquad i = 1, 2, \cdots, M \qquad (6.40)$$

where

$$F_i(\gamma_i(k+1)) = \frac{exp[-\frac{1}{2}\gamma_i^T(k+1)V_i^{-1}(k)\gamma_i(k+1)]}{(2\pi)^{\frac{m-1}{2}} det(V_i(k))]^{\frac{1}{2}}} \qquad (6.41)$$

$$V_i(k) = E(\gamma_i(k)\gamma_i^T(k)) = \hat{C}_i(k)P_i(k|k-1)\hat{C}_i^T(k) + R_i(k) \qquad (6.42)$$

There is the difference between equations (6.24)-(6.26) and equations (6.40)-(6.42). In equation (6.42), $\hat{C}_i$ and R_i instead of C_i and R are used. Here R_i is the covariance of the measurement noise v_i. In equation (6.40), the sum is from $j = 1$ to M rather than from $j = 0$ to M. In equation (6.41), $(2\pi)^{\frac{m-1}{2}}$, instead of $(2\pi)^{\frac{m}{2}}$, is used.

The algorithm for fault detection and state estimation is expressed as follows:

1. For some $i \in \{1, 2, \cdots, M\}$, if the following conditions hold

$$p_i(k) \gg p_j(k) \qquad \forall j \neq i \qquad (6.43)$$

$$Max\{p_1(k), \cdots, p_{i-1}(k), p_{i+1}(k), \cdots, p_M(k)\} < \varepsilon \qquad (6.44)$$

then the ith sensor has failed. The final state estimate is

$$\hat{x}(k) = \hat{x}_i(k) \qquad (6.45)$$

2. If there is no $i \in \{1, 2, \cdots, M\}$ which satisfies the conditions (6.43) and (6.44), then no sensor fails. The system works under the normal condition. The final state estimate is

$$\hat{x}(k) = \hat{x}_0(k) \qquad (6.46)$$

The $p_i(k)$ expressed in equation (6.40) may not be exactly the probability that the ith unreliable sensor has failed since the M Kalman filters use different measurement data. However, this scheme does reduce the computational complexity of state estimate and failure detection. It is preferable in practical application. The scheme can also be modified directly to the multiple-failure case by applying the results obtained in Scheme 2.

To further reduce the real-time computational burden, we can use the steady state Kalman filters for time-invariant systems. In a steady state Kalman filter, $K_i(k)$ and $V_i^{-1}(k)$ are both constant matrices and can be calculated off-line. In this way, the real-time computational burden can be reduced to a very moderate level.

One of the most significant advantages of this technique is its inherent ability for state estimator reorganization subsequent to the occurrence of sensor failure.

6.2.3   Application to the presssurized headbox

To achieve the fault-tolerant control, another important issue is the system reorganization to retain system integrity in the presence of failures. In the previous section, three fault detection and state estimator reorganization schemes have been developed. In this section, we apply Scheme 3 to the design of fault-tolerant control system for the pressurized paper machine headbox.

The system dynamics and measurement equation of the pressurized headbox have been given in equations (6.9)-(6.11). As mentioned above, all the three sensors may fail during operation. There will be, therefore, four hypotheses to be considered, as follows

$$H_0: \quad y(k) = \begin{pmatrix} 1.0 & 0.0 \\ 0.0 & 1.0 \\ 1.0 & -1.0 \end{pmatrix} x(k) + v(k) \tag{6.47}$$

$$H_1: \quad y(k) = \begin{pmatrix} 0.0 & 0.0 \\ 0.0 & 1.0 \\ 1.0 & -1.0 \end{pmatrix} x(k) + d_1(k) + v(k) \tag{6.48}$$

$$H_2: \quad y(k) = \begin{pmatrix} 1.0 & 0.0 \\ 0.0 & 0.0 \\ 1.0 & -1.0 \end{pmatrix} x(k) + d_2(k) + v(k) \tag{6.49}$$

$$H_3: \quad y(k) = \begin{pmatrix} 1.0 & 0.0 \\ 0.0 & 1.0 \\ 0.0 & 0.0 \end{pmatrix} x(k) + d_3(k) + v(k) \tag{6.50}$$

where H_0 is the hypothesis for normal condition. H_1, H_2 and H_3 are the hypotheses which assume the total head sensor, stock level sensor and air pressure sensor have failed, respectively. For each hypothesis, the system is completely observable.

The selection of the variance of system noise $w(k)$ should consider the uncertainty of our knowledge in the process dynamics, the relative scale of the variables and environment disturbances. The selection of the variance of measurement noise $v(k)$ should consider the accuracy of the sensors. Through careful investigation, we select $Q = diag\,(0.025^2, 0.015^2)$, $R = diag\,(0.035^2, 0.02^2, 0.05^2)$.

To reduce the real-time computational burden, four steady state Kalman filters, denoted by $KF_0 \sim KF_3$, can be designed under different failure assumptions. $p_1(k)$, $p_2(k)$ and $p_3(k)$ are calculated by equations (6.40)-(6.42). If stock level sensor fails, both $p_1(k)$ and $p_3(k)$ will be less than ε. As a result, $p_2(k)$ will be large than $1 - 2\varepsilon$. Similar occasions will occur if total head sensor or air pressure sensor fails.

The total head H is the summation of the stock level L and the air pressure P. Since the three sensors have the same scale, there is a direct relation among the outputs of the sensors

$$y^1(k) = y^2(k) - y^3(k) \tag{6.51}$$

where y^i (i=1, 2, 3) is the ith component of the measurement vector y. When a sensor fails, the above relation will not hold. Thus, equation (6.51) can be applied to fault detection.

According to the discussion above and selecting $\varepsilon = 0.005$, the fault detection and system reorganization use the following mechanism:

1. The normal unfailed condition

If $|y_1(k) - y_2(k) - y_3(k)| \leq 0.15$ or $Max\{p_1(k), p_2(k), p_3(k)\} \leq 0.99$, then the system is in the normal condition. The final state estimate is

$$\hat{x}(k) = \hat{x}_0(k) \tag{6.52}$$

2. The total head sensor failure

If $|y_1(k) - y_2(k) - y_3(k)| > 0.15$ and $p_1(k) > 0.99$, then the total head sensor has failed. The final state estimate is

$$\hat{x}(k) = \hat{x}_1(k) \tag{6.53}$$

3. The stock level sensor failure

If $|y_1(k) - y_2(k) - y_3(k)| > 0.15$ and $p_2(k) > 0.99$, then the stock level sensor has failed. The final state estimate is

$$\hat{x}(k) = \hat{x}_2(k) \tag{6.54}$$

4. The air pressure sensor failure

If $|y_1(k) - y_2(k) - y_3(k)| > 0.15$ and $p_3(k) > 0.99$, then the air pressure sensor has failed. The final state estimate is

$$\hat{x}(k) = \hat{x}_3(k) \tag{6.55}$$

The decoupling control law is realized by applying the final state estimate

$$u(k) = F\hat{x}(k) + Gv_m(k) \tag{6.56}$$

where the feedback gain F and the feedforward gain G are synthesized using the state feedback decoupling algorithm proposed by Gilbert (1969)

$$F = \begin{pmatrix} 0.498 & -0.291 \\ 0.402 & 2.45 \end{pmatrix} \quad G = \begin{pmatrix} 0.0731 & 0.260 \\ 0.247 & -3.20 \end{pmatrix}$$

For performance comparison, a conventional decoupling controller is also implemented by applying the state estimate of the Kalman filter under normal unfailed assumption

$$u(k) = F\hat{x}_0(k) + Gv_m(k) \tag{6.57}$$

Note that in this section, the failure detection conditions (6.43) and (6.44) have been reduced to the simpler approximate form, $p_i > 1 - (M - 1)\epsilon$, for convenience.

From (6.40), it can be seen that if $p_i(k)$ is small, $p_i(k+1)$ will grow very slowly at. In order to avoid this drastic effect, a lower bound is set on $p_i(k)$. Here we select 0.001 as the lower bound. Besides, when a sensor failure occurs, the Kalman filter based on the wrong hypothesis may diverge. It will take a long time before the estimation error decreased to reflect the trueness of the hypothesis after the sensor has been repaired. To solve the problem, we reset the estimate of potentially divergent Kalman filter to the final state estimate.

The performance of the headbox fault-tolerant control system has been studied using digital simulations. The initial state is assumed to be zero. The setpoint of total head is increased to 1 at k=150 and then goes back to 0 at k=350. The setpoint of stock level is increased to 1 at k=50 and then goes back to 0 at k=250. It is assumed that failures occur at k=100 and the failed sensors have a constant output, 0.2. Figures 6.1 and 6.2 show the responses of the fault-tolerant control system and conventional decoupling control system, respectively, when the total head sensor has failed; Figures 6.3 and 6.4 show those when the stock level sensor has failed; Figures 6.5 and 6.6 show those when the air pressure sensor has failed. The solid curves represent the setpoint and actual response of the stock level. The doted curves represent that of the total head. From these figures, it is obvious that when any sensor has failed, the dynamic performance of the conventional decoupling control system becomes very poor and the system is no longer decoupled. However, the fault tolerant control system still remains satisfactorily dynamic and decoupling performances. Thus, the efficiency of the proposed control technique to deal with sensor failures has been proven.

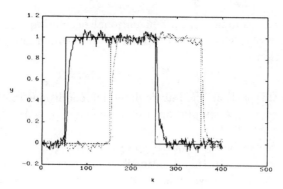

Figure 6.1: Response of fault-tolerant control when total head sensor fails

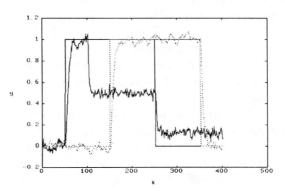

Figure 6.2: Response of conventional decoupling control when total head sensor fails

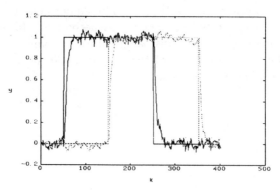

Figure 6.3: Response of fault-tolerant control when stock level sensor fails

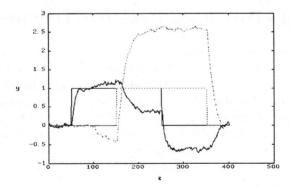

Figure 6.4: Response of conventional decoupling control when stock level sensor fails

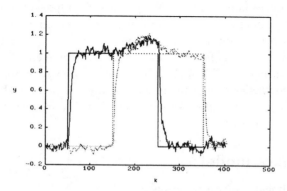

Figure 6.5: Response of fault-tolerant control when air pressure sensor fails

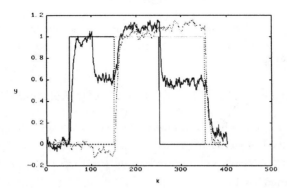

Figure 6.6: Response of conventional decoupling control when air pressure sensor fails

6.3   Fault-Tolerance Design of Drying Section

Linear quadratic optimal control technique provides a straightforward means for synthesizing stable linear feedback systems. It has been applied widely to industrial processes. The robustness and fault-tolerance of control systems are extremely crucial in the industrial application of linear quadratic optimal control technique. There are a great number of researches on the robustness of linear quadratic regulators (Vidyasagar and Viswandaham, 1985; Patel, et al. 1977; Yedavalli, 1985). However, few have been done for the fault-tolerance of the systems. The control systems are designed under the assumption that components do not fail.

There are a number of researches on fault-tolerant control using the methods other than linear quadratic optimal control. Those researches investigated complete component faults (Siljak, 1980). Once a component has failed, it will not function any more and should be discarded from the control system. In industrial processes, however, actuators and sensors usually suffer performance degradation. It is reasonable to assume that the components fail partially, or degrade in performance. This section investigates fault-tolerance analysis and fault-tolerant controller design for linear quadratic system with performance degradation of actuators and sensors, as well as their application to the drying section of a paper machine.

6.3.1   Failure model

Consider a linear multivariable system

$$\dot{x}(t) = Ax(t) + Bu(t) \tag{6.58}$$

where $x \in \Re^n$, $u \in \Re^m$ are state and control vectors, respectively; A, B are $n \times n$ and $n \times m$ constant matrices. The performance criterion with the constraint (6.58) is

$$J = \frac{1}{2} \int_0^\infty e^{2\sigma t}(x^T(t)\rho Q_0 x(t) + u^T(t)Ru(t))\, dt \tag{6.59}$$

where $\sigma \geq 0$ is a scalar which represents the specified degree of stability, $\rho > 0$ is a scalar involved in weighting matrix, Q_0 and R are the symmetric positive semi-definite and positive definite weighting matrices respectively. The pair (A, B) is assumed to be controllable and $(A, Q_0^{\frac{1}{2}})$ to be observable. The optimal control law is given by (Anderson and Moore, 1971) as

$$u(t) = Kx(t) \tag{6.60}$$

with the optimal control gain

$$K = -R^{-1}B^T P \tag{6.61}$$

where P satisfies

$$A^T P + PA + 2\sigma P + \rho Q_0 - PBR^{-1}B^T P = 0 \tag{6.62}$$

The closed-loop system is of the form

$$\dot{x}(t) = A_c x(t) \tag{6.63}$$

where

$$A_c = A + BK \tag{6.64}$$

It is assumed that the actuator and sensor faults have two effects:

(1) the actuator and sensor gains may be reduced from its nominal value unit to zero;

(2) the actuator action and sensor output may contain a bias.

These two effects can be described by

$$\tilde{u}_i(t) = \alpha_i u_i(t) + u_{i0} \tag{6.65}$$

$$\tilde{x}_j(t) = \beta_j x_j(t) + x_{j0} \tag{6.66}$$

where u_i and $\tilde{u}_i$ are the normal and fault control actions of the ith actuator, x_j and $\tilde{x}_j$ are the normal and fault outputs of the jth sensor, u_{i0} and x_{j0} are biases, and $0 \leq \alpha_i \leq 1$, $0 \leq \beta_j \leq 1$ are gain degradation factors of the ith actuator and the jth sensor. $\alpha_i = 1$, $\alpha_i = 0$ and $0 < \alpha_i < 1$ represent the normal condition, complete failure and partial failure (performance degradation) of the ith actuator, respectively.

As far as the eigenvalue location is concerned for stability analysis, only the multiplicative effect is important. The fault degree of the ith actuator and jth sensor can be defined as $1 - \alpha_i$ and $1 - \beta_j$, respectively. Therefore, the actuator fault is equivalent to reducing the value in the corresponding row of the feedback gain matrix. Whereas the sensor fault is equivalent to reducing the value in the corresponding column of the feedback gain matrix.

When the ith actuator fails, the closed-loop dynamics becomes

$$
\begin{aligned}
\dot{x}(t) &= Ax(t) + B(u_1(t) \cdots \alpha_i u_i(t) \cdots u_m(t))^T \\
&= A_c x(t) - (1 - \alpha_i) E_i^a x(t)
\end{aligned}
\tag{6.67}
$$

where E_i^a is the influence matrix of the ith actuator fault

$$E_i^a = B_i^c K_i^r \tag{6.68}$$

with B_i^c, the ith column of matrix B, and K_i^r, the ith row of feedback gain K. Similarly, when the jth sensor fails

$$\dot{x}(t) = A_c x(t) - (1 - \beta_j) E_j^s x(t) \tag{6.69}$$

where E_j^s is the influence matrix of the jth sensor given by

$$E_j^s = (0 \ \cdots \ BK_j^c \ \cdots \ 0) \tag{6.70}$$

with K_j^c, the jth column of feedback gain K.

We assume that there are m' $(\leq m)$ actuators and n' $(\leq n)$ sensors which are not reliable and tend to fail. Denoting $\Phi = \{a_1, a_2, \cdots, a_{m'}\} \subset \{1, 2, \cdots, m\}$ and $\Psi = \{s_1, s_2, \cdots, s_{n'}\} \subset \{1, 2, \cdots, n\}$ as the sets of unreliable actuators and sensors, respectively, a general fault model of the closed-loop control system follows

$$\dot{x}(t) = A_c x(t) - \left(\sum_{i \in \Phi} (1 - \alpha_i) E_i^a + \sum_{j \in \Psi} (1 - \beta_j) E_j^s \right) x(t) \tag{6.71}$$

According to the discussion above, the fault-tolerance of linear quadratic systems can be defined as the bound of permitted sensor and actuator gain degradations to maintain system (6.71) to be stable.

6.3.2   Fault-tolerance analysis

According to the definition of fault-tolerance in the section 6.2, the problem of concern in fault-tolerance analysis is to obtain the conditions on the magnitude of $(1 - \alpha_i)$ and $(1 - \beta_j)$ such that the closed loop system remains stable. For this purpose, we will reveal under what conditions the quadratic function $V(x) = x^T(t) P x(t)$ is a Lyapunov function for the fault model (6.71). The quantitative results concerning the fault-tolerance are obtained by applying Lyapunov's stability theorem.

Theorem 1 *The fault-tolerance of the linear quadratic system (6.58)-(6.59) is given by*

$$\left(\sum_{i \in \Phi} (1 - \alpha_i)^2 + \sum_{j \in \Psi} (1 - \beta_j)^2 \right)^{\frac{1}{2}} < \mu \tag{6.72}$$

with

$$\mu = \lambda_{min}(D) / \parallel P^e \parallel_s \tag{6.73}$$

or

$$\sum_{i\in\Phi}(1-\alpha_i)\parallel P_i^a \parallel_s + \sum_{j\in\Psi}(1-\beta_j)\parallel P_j^s \parallel_s < \lambda_{min}(D) \qquad (6.74)$$

where $\parallel P^e \parallel_s = \lambda_{max}^{\frac{1}{2}}(P^e(P^e)^T)$ *is the spectral norm,* $\lambda_{min}(D)$ *and* $\lambda_{max}(D)$
are the smallest and largest eigenvalue of the matrices respectively, and

$$P_i^a = (E_i^a)^T P + P E_i^a, \qquad i\in\Phi \qquad (6.75)$$

$$P_j^s = (E_j^s)^T P + P E_j^s, \qquad j\in\Psi \qquad (6.76)$$

$$D = \rho Q_0 + PBR^{-1}B^T P + 2\sigma P \qquad (6.77)$$

$$P^e = (P_{a_1}^a \quad \cdots \quad P_{a_{m'}}^a, \ P_{s_1}^s \quad \cdots \quad P_{s_{n'}}^s) \qquad (6.78)$$

Proof: Define the Lyapunov function as

$$V(x) = x^T(t)Px(t) \qquad (6.79)$$

Taking derivative on both sides of equation (6.79) gives

$$\dot{V}(t) = \dot{x}^T(t)Px(t) + x^T(t)P\dot{x}(t) \qquad (6.80)$$

Substituting equation (6.71) into (6.80), we obtain

$$\begin{aligned}
\dot{V}(x) = \ & x^T(t)(A_c^T P + PA_c)x(t) \\
& -x^T(t)[(\sum_{i\in\Phi}(1-\alpha_i)E_i^a + \sum_{j\in\Psi}(1-\beta_j)E_j^s)^T P \\
& +P(\sum_{i\in\Phi}(1-\alpha_i)E_i^a + \sum_{j\in\Psi}(1-\beta_j)E_j^s)]x(t)
\end{aligned} \qquad (6.81)$$

With (6.62), equation (6.81) can be rewritten as

$$\begin{aligned}
\dot{x}(t) = \ & -x^T(t)(\rho Q_0 + PBR^{-1}B^T P + 2\sigma P)x(t) \\
& +x^T(t)[\sum_{i\in\Phi}(\alpha_i-1)P_i^a + \sum_{j\in\Psi}(\beta_j-1)P_j^s]x(t) \\
= \ & -x^T(t)Dx(t) + x^T(t)P^g x(t)
\end{aligned} \qquad (6.82)$$

where D is given in equation (6.77), and

$$P^g = \sum_{i\in\Phi}(\alpha_i-1)P_i^a + \sum_{j\in\Psi}(\beta_j-1)P_j^s \qquad (6.83)$$

From equation (6.82), it is obvious that if

$$x^T(t)Dx(t) \geq x^T(t)P^g x(t) \tag{6.84}$$

then $\dot{V}(x) \leq 0$, the closed-loop system is stable.

Since

$$x^T(t)Dx(t) \geq \lambda_{min}(D)x^T(t)x(t) \tag{6.85}$$

and

$$x^T(t)P^g x(t) \leq \| P^g \|_s x^T(t)x(t) \tag{6.86}$$

then the following condition must be hold in order to guarantee a stable closed-loop system

$$\| P^g \|_s \leq \lambda_{min}(D) \tag{6.87}$$

Note that

$$\| P^g \|_s \leq \sum_{i \in \Phi}(1 - \alpha_i) \| P_i^a \|_s + \sum_{j \in \Psi}(1 - \beta_j) \| P_j^s \|_s \tag{6.88}$$

It is obvious that equation (6.87) implies equation (6.74).

Similarly, if we rewrite P^g into

$$P^g = P^e[(\alpha_1 - 1)I \cdots (\alpha_m - 1)I \ (\beta_1 - 1)I \cdots (\beta_n - 1)I] \tag{6.89}$$

where P^e is given in (6.78), then

$$\| P^g \|_s \leq (\sum_{i \in \Phi}(\alpha_i - 1)^2 + \sum_{j \in \Psi}(\beta_j - 1)^2)^{\frac{1}{2}} \| P^e \|_s \tag{6.90}$$

Equation (6.72) follows directly from (6.87).

Q.E.D.

The above theorem considers general situations of fault-tolerance. In real-world industrial systems, it is reasonable to assume that only single fault is possible. We have the following lemma for single fault case from Theorem 1.

Lemma 1 *The fault-tolerance against the ith actuator and jth sensor is given, respectively, by*

$$1 - \alpha_i > \frac{\lambda_{min}(D)}{\| P_i^a \|_s} \tag{6.91}$$

$$1 - \beta_j > \frac{\lambda_{min}(D)}{\| P_j^s \|_s} \tag{6.92}$$

Theorem 1 provides the fault-tolerance for a given stabilizing optimal controller. The following theorem relates the fault-tolerance to the dominant eigenvalues of the closed-loop system.

Theorem 2 *The fault-tolerance is proportional to the real part of the dominant eigenvalues of nominal closed-loop system*

$$\mu \propto -Re(\lambda_{max}(A_c)) \qquad (6.93)$$

where P and μ are given in (6.62) and (6.73), respectively.

Proof: To simplify the proof, we only consider the single fault case, i.e., the ith actuator fault. The results obtained are also applicable to the general situation.

From equations (6.62) and (6.77), we have

$$A_c^T P + P A_c = -D \qquad (6.94)$$

Denoting $\lambda_j(A_c)$ and $v_j(A_c)$ as the eigenvalues and corresponding eigenvectors of A_c, we obtain

$$A_c v_j(A_c) = \lambda_j(A_c) v_j(A_c) \qquad j = 1, 2, \cdots, n \qquad (6.95)$$

Premultiplying $v_j^*(A_c)$ and postmultiplying $v_j(A_c)$ in both sides of equation (6.94) gives

$$
\begin{aligned}
& v_j^*(A_c)(A_c^T P + P A_c) v_j(A_c) \\
= {} & (\lambda_j^*(A_c) + \lambda_j(A_c)) v_j^*(A_c) P v_j(A_c) \\
= {} & 2Re(\lambda_j(A_c)) v_j^*(A_c) P v_j(A_c) \\
= {} & -v_j^*(A_c) D v_j(A_c) \qquad (6.96)
\end{aligned}
$$

where λ_j^* and v_j^* are the conjugate transpose of λ_j and v_j. Thus

$$-2Re(\lambda_j(A_c)) = \frac{v_j^*(A_c) D v_j(A_c)}{v_j^*(A_c) P v_j(A_c)} \qquad (6.97)$$

Similarly, premultiplying $v_j^*(E_i^a)$ and postmultiplying $v_j(E_i^a)$ in both sides of equation (6.75), we have

$$-2Re(\lambda_j(E_i^a)) = \frac{v_j^*(E_i^a) P_i^a v_j(E_i^a)}{v_j^*(E_i^a) P v_j(E_i^a)} \qquad (6.98)$$

Since P_i^a, P and D are all symmetric matrices, by applying Rayleigh theorem the following inequalities are obtained

$$\lambda_{min}(D) \leq \frac{v_j^*(A_c)Dv_j(A_c)}{v_j^*(A_c)v_j(A_c)} \leq \lambda_{max}(D) \tag{6.99}$$

$$\lambda_{min}(P) \leq \frac{v_j^*(A_c)Pv_j(A_c)}{v_j^*(A_c)v_j(A_c)} \leq \lambda_{max}(P) \tag{6.100}$$

$$\lambda_{min}(P) \leq \frac{v_j^*(E_i^a)Pv_j(E_i^a)}{v_j^*(E_i^a)v_j(E_i^a)} \leq \lambda_{max}(P) \tag{6.101}$$

$$\lambda_{min}(P_i^a) \leq \frac{v_j^*(E_i^a)P_i^a v_j(E_i^a)}{v_j^*(E_i^a)v_j(E_i^a)} \leq \lambda_{max}(P_i^a) \tag{6.102}$$

Since

$$\frac{\lambda_{min}(D)}{\| P_i^a \|_s} \leq \frac{\lambda_{min}(D)\,\lambda_{min}(P)}{\lambda_{max}(P)\,\| P_i^a \|_s} \tag{6.103}$$

and according to the property of singular values of matrices, we have

$$\| P_i^a \|_s = \lambda_{max}^{\frac{1}{2}}(P_i^a P_i^a) = max|\lambda(P_i^a)| \tag{6.104}$$

Substitute equations (6.99) through (6.102) and (6.104) into (6.103) yields

$$\begin{aligned}
\frac{\lambda_{min}(D)}{\| P_i^a \|_s} \leq\ & [\frac{v_j^*(A_c)Dv_j(A_c)}{v_j^*(A_c)v_j(A_c)}/(\frac{v_j^*(A_c)Pv_j(A_c)}{v_j^*(A_c)v_j(A_c)})] \\
\times\ & [\frac{v_j^*(E_i^a)Pv_j(E_i^a)}{v_j^*(E_i^a)v_j(E_i^a)}/(\frac{|v_j^*(E_i^a)P_i^a v_j(E_i^a)|}{v_j^*(E_i^a)v_j(E_i^a)})] \\
=\ & \frac{v_j^*(A_c)Dv_j(A_c)}{v_j^*(A_c)Pv_j(A_c)}\frac{v_j^*(E_i^a)Pv_j(E_i^a)}{|v_j^*(A_c)P_i^a v_j(A_c)|}
\end{aligned} \tag{6.105}$$

Applying equations (6.97) and (6.98) to (6.105) yields

$$\mu = \frac{\lambda_{min}(D)}{\| P_i^a \|_s} \leq -\frac{Re(\lambda_{max}(A_c))}{min|Re(\lambda(E_i^a))|} \tag{6.106}$$

Q.E.D.

Since increasing σ and ρ will increase $-Re(\lambda_{max}(A_c))$ (Patel, et al. 1977), larger σ and ρ will result in higher fault-tolerance. However, increasing σ and ρ will also increase P and $min|Re(\lambda(E_i^a))|$. These two reverse effects imply that there exist optimal values for σ and ρ which give the highest fault-tolerance.

6.3.3 Fault-tolerance design

Based on equations (6.72) and (6.73), we define a scalar J as

$$J = 1/\mu = \frac{\| P^e \|_s}{\lambda_{min}(D)} \tag{6.107}$$

The absolute maximum of fault-tolerance implies the absolute minimum of J. Therefore, the problem of designing a controller with the highest fault-tolerance can be formulated by: finding optimal values of σ and ρ to minimize J

$$Min_{\sigma,\rho}\{J\} = Min_{\sigma,\rho}\{\frac{\| P^e \|_s}{\lambda_{min}(D)}\} \tag{6.108}$$

In order to implement a gradient-based descent procedure, we derive the gradient of J with respect to σ and ρ.

Theorem 3 *The gradient of J with respect to σ is described by*

$$\frac{\partial J}{\partial \sigma} = \frac{1}{\lambda_{min}^2(D)}(\frac{1}{2 \| P^e \|_s}\lambda_{min}(D)Tr\{(u_n v_n^T + v_n u_n^T)P^e L_\sigma^e\}$$
$$- \| P^e \|_s Tr\{(v_d u_d^T + u_d v_d^T)(-M_\sigma BK + P + \sigma M_\sigma)\}) \tag{6.109}$$

where v_n and u_n are the right and left singular vectors corresponding to $\lambda_{max}(P^e(P^e)^T)$, v_d and u_d are that corresponding to $\lambda_{min}(D)$, respectively, M_σ satisfies

$$A_p^T M_\sigma + M_\sigma A_p + 2P = 0 \tag{6.110}$$

and

$$L_\sigma^e = (L_{\sigma a_1}^a \cdots L_{\sigma a_m}^a, L_{\sigma s_1}^s \cdots L_{\sigma s_{n'}}^s) \tag{6.111}$$

$$L_{\sigma i}^a = (N_{\sigma i}^a)^T P + P N_{\sigma i}^a + (E_i^a)^T M_\sigma + M_\sigma E_i^a \quad i \in \Phi \tag{6.112}$$

$$L_{\sigma j}^s = (N_{\sigma j}^s)^T P + P N_{\sigma j}^s + (E_j^s)^T M_\sigma + M_\sigma E_j^s \quad j \in \Psi \tag{6.113}$$

$$N_{\sigma i}^a = B_i^c K_{\sigma i}^\tau, \quad N_{\sigma j}^s = (0 \cdots BK_{\sigma j}^c \cdots 0) \tag{6.114}$$

$$K_\sigma = -R^{-1} B^T M_\sigma, \quad A_p = A_c + \sigma I \tag{6.115}$$

The gradient of J with respect to ρ is described by

$$\frac{\partial J}{\partial \rho} = \frac{1}{\lambda_{min}^2(D)}(\frac{1}{2 \| P^e \|_s}\lambda_{min}(D)Tr\{(u_n v_n^T + v_n u_n^T)P^e L_\rho^e\}$$
$$- \| P^e \|_s Tr\{(v_d u_d^T + u_d v_d^T)(-M_\rho BK + \frac{Q_0}{2} + \sigma M_\rho)\}) \tag{6.116}$$

where M_ρ satisfies

$$A_p^T M_\rho + M_\rho A_p + Q_0 = 0 \qquad (6.117)$$

Other matrices in equation (6.116) have the same forms as those for the gradient with respect to σ.

Proof: From equation (6.107), we have

$$J = \frac{Tr\{\| P^e \|_s\}}{Tr\{\lambda_{min}(D)\}} \qquad (6.118)$$

From equation (6.118), the deviation of J is obvious

$$\begin{aligned}
\Delta J &= \frac{1}{\lambda_{min}^2(D)} \big(\frac{1}{2\| P_e \|_s} \lambda_{min}(D) Tr\{\Delta \lambda_{max}(P^e P^{eT})\} \\
&\quad - \| P^e \|_s Tr\{\Delta \lambda_{min}(D)\} \big)
\end{aligned} \qquad (6.119)$$

where

$$\begin{aligned}
\Delta \lambda_{max}(P^e P^{eT}) &= v_n u_n^T \Delta(P^e P^{eT}) \\
&= v_n u_n^T (\Delta P^e P^{eT} + P^e \Delta P^{eT}) \qquad (6.120)
\end{aligned}$$

$$\Delta \lambda_{min}(D) = v_d u_d^T \Delta D \qquad (6.121)$$

where v_d and u_d are the right and left singular vectors corresponding to $\lambda_{max}(P^e(P^e)^T)$, v_n and u_n are the right and left singular vectors corresponding to $\lambda_{min}(D)$, respectively.

From equations (6.61), (6.68), (6.70) and (6.75) through (6.78), it is obvious that

$$\Delta P^e = (\Delta P_{a_1}^a \quad \cdots \quad \Delta P_{a_m'}^a, \ \Delta P_{s_1}^s \quad \cdots \quad \Delta P_{s_{n'}}^s) \qquad (6.122)$$

$$\Delta P_i^a = (\Delta E_i^a)^T P + (E_i^a)^T \Delta P + \Delta P E_i^a + P \Delta E_i^a \quad i \in \Phi \qquad (6.123)$$

$$\Delta P_j^s = (\Delta E_j^s)^T P + (E_j^s)^T \Delta P + \Delta P E_j^s + P \Delta E_j^s \quad j \in \Psi \qquad (6.124)$$

with

$$\Delta E_i^a = B_i^c \Delta K_i^r \qquad (6.125)$$

$$\Delta E_j^s = (0 \quad \cdots \quad B \Delta K_j^c \cdots \quad 0) \qquad (6.126)$$

$$\Delta K = -R^{-1} B^T \Delta P \qquad (6.127)$$

Taking the perturbation in the Riccati equation (6.62) with respect to σ, we have

$$\begin{aligned}
& A^T (P + \Delta P_\sigma) + (P + \Delta P_\sigma)A + 2(\sigma + \Delta \sigma)(P + \Delta P_\sigma) \\
&+ \rho Q_0 - (P + \Delta P_\sigma)BR^{-1}B^T(P + \Delta P_\sigma) = 0
\end{aligned} \qquad (6.128)$$

Neglecting the $O(\Delta^2)$ terms gives

$$A_p^T \Delta P_\sigma + \Delta P_\sigma A_p = -2\Delta\sigma P \qquad (6.129)$$

where A_p is given in equation (6.116). The solution of (6.129) is of the form (Desouze and Bhattacharyya, 1981)

$$\Delta P_\sigma = M_\sigma \Delta\sigma \qquad (6.130)$$

where

$$M_\sigma = 2 \sum_{i=1}^{n} \sum_{k=1}^{n} \gamma_{ik} (A_p^T)^{i-1} P (A_p)^{k-1} \qquad (6.131)$$

Thus, M_σ satisfies the following equation

$$A_p^T M_\sigma + M_\sigma A_p + 2P = 0 \qquad (6.132)$$

The resulted perturbation of control gain K is

$$\Delta K_\sigma = -R^{-1} B^T M_\sigma \Delta\sigma \qquad (6.133)$$

Substituting (6.130) and (6.133) into (6.123)-(6.126), we get

$$\Delta E_i^a = N_{\sigma i}^a \Delta\sigma \qquad (6.134)$$

$$\Delta E_j^s = N_{\sigma j}^s \Delta\sigma \qquad (6.135)$$

$$\Delta P_{\sigma i}^a = L_{\sigma i}^a \Delta\sigma \qquad (6.136)$$

$$\Delta P_{\sigma j}^s = L_{\sigma j}^s \Delta\sigma \qquad (6.137)$$

where

$$N_{\sigma i}^a = -B_i^c (R^{-1} B^T M_\sigma)_i^r \qquad (6.138)$$

$$N_{\sigma j}^s = (0 \quad \cdots \quad -B(R^{-1} B^T M_\sigma)_j^c \quad \cdots \quad 0) \qquad (6.139)$$

$$L_{\sigma i}^a = (N_{\sigma i}^a)^T P + (E_i^a)^T M_\sigma + M_\sigma E_i^a + P N_{\sigma i}^a \qquad (6.140)$$

$$L_{\sigma j}^s = (N_{\sigma j}^s)^T P + (E_j^s)^T M_\sigma + M_\sigma E_j^s + P N_{\sigma j}^s \qquad (6.141)$$

From equation (6.77), we obtain

$$\begin{aligned} \Delta D_\sigma &= \Delta P B R^{-1} B^T P + P B R^{-1} B^T \Delta P + 2\sigma\Delta P + 2P\Delta\sigma \\ &= (M_\sigma B R^{-1} B^T P + P B R^{-1} B^T M_\sigma + 2P + 2\sigma M_\sigma)\Delta\sigma \end{aligned} \qquad (6.142)$$

Since

$$Tr\{\Delta\lambda_{max}(P^e P^{eT})\} = Tr\{(u_n v_n^T + v_n u_n^T)P^e \Delta P^{eT}\} \qquad (6.143)$$

and from equations (6.122), (6.136) and (6.137) we have

$$\Delta P^e = (L^a_{\sigma a_1} \cdots L^a_{\sigma a_{m'}} \cdots L^s_{\sigma s_{m'}})\Delta\sigma \tag{6.144}$$

Thus

$$Tr\{\Delta\lambda_{max}(P^e P^{eT})\} = Tr\{(u_n v_n^T + v_n u_n^T)P^e L^e\}\Delta\sigma \tag{6.145}$$

Substituting equations (6.120), (6.121), (6.142) and (6.145) into (6.119) gives

$$
\begin{aligned}
\Delta J_\sigma = & \; [\frac{1}{\lambda^2_{min}(D)}(\frac{1}{2\parallel P^e \parallel_s}\lambda_{min}(D) * Tr\{(u_n v_n^T + v_n u_n^T)P^e L^e_\sigma\} \\
& - \parallel P^e \parallel_s Tr\{(v_d u_d^T + u_d v_d^T) \\
& (-M_\sigma BK + P + \sigma M_\sigma)\})]\Delta\sigma
\end{aligned}
\tag{6.146}
$$

Taking perturbation in the Riccati equation (6.62) with respect to ρ and manipulation in the similar way as above, it is easy to obtain

$$
\begin{aligned}
\Delta J_\rho = & \; \frac{1}{\lambda^2_{min}(D)}(\frac{1}{2\parallel P^e \parallel_s}\lambda_{min}(D)Tr\{(u_n v_n^T + v_n u_n^T)P^e L^e_\rho\} \\
& - \parallel P^e \parallel_s Tr\{(v_d u_d^T + u_d v_d^T) \\
& (-M_\rho BK + \frac{Q_0}{2} + \sigma M_\rho)\})\Delta\rho
\end{aligned}
\tag{6.147}
$$

where M_ρ satisfies

$$A_p^T M_\rho + M_\rho A_p + Q_0 = 0 \tag{6.148}$$

and

$$L^e_\rho = (L^a_{\rho a_1} \cdots L^a_{\rho a_{m'}} L^s_{\rho s_1} \cdots L^s_{\rho s_{n'}}) \tag{6.149}$$

$$L^a_{\rho i} = (N^a_{\rho i})^T P + PN^a_{\rho i} + (E^a_i)^T M_\rho + M_\rho E^a_i \tag{6.150}$$

$$L^s_{\rho j} = (N^s_{\rho j})^T P + PN^s_{\rho j} + (E^s_j)^T M_\rho + M_\rho E^s_j \tag{6.151}$$

$$N^a_{\rho i} = B^c_i K^r_{\rho i}, \quad N^s_{\rho j} = (0 \cdots BK^c_{\rho i} \cdots 0) \tag{6.152}$$

$$K_\rho = R^{-1}B^T M_\rho, \quad A_p = A_c + \sigma I \tag{6.153}$$

From equations (6.146) and (6.147), equations (6.109) and (6.116) is proven.

$$\text{Q.E.D.}$$

With Theorem 3, the algorithm which improves fault-tolerance from selecting optimal σ and ρ can be easily designed using the gradient-based method. To improve the convergency and efficiency of the iteration process, Armijo algorithm (Polak, 1971) can be applied to choose step size.

We define the controller parameter vector λ as

$$\lambda = \begin{pmatrix} \sigma \\ \rho \end{pmatrix} \tag{6.154}$$

thus

$$\nabla J(\lambda) = \begin{pmatrix} \frac{\partial J}{\partial \sigma} \\ \frac{\partial J}{\partial \rho} \end{pmatrix} \tag{6.155}$$

Introducing a 2×2 positive definite matrix $G(\lambda)$ and a set

$$C(\lambda_0) = \{\lambda | J(\lambda) \leq J(\lambda_0)\} \tag{6.156}$$

and a function

$$\theta(\varphi; \lambda) = [J(\lambda + \varphi h(\lambda)) - J(\lambda)] - \varphi \xi \langle \nabla J(\lambda), h(\lambda) \rangle \tag{6.157}$$

where $h(\lambda)$ is given below, "$\langle \cdot \rangle$" represents point product. The gradient-based algorithm can be stated as follows:

Step 1 Select λ_0 such that the set $C(\lambda_0)$ is bounded; select $\xi \in (0, 1)$, $\eta \in (0, 1)$ and $\omega > 0$. $\lambda_0 = 1$, $\xi = \frac{1}{2}$, $\eta \in (0.5, 0.8)$ and $\omega = 1$ are recommended;

Step 2 Set $l = 0$;

Step 3 Compute $h(\lambda_l) = -G(\lambda_l) \nabla J(\lambda_l)$ here $G(\lambda_l) = I$ is selected;

Step 4 If $|h(\lambda^l)| \leq \epsilon$ (ϵ is a small positive scalar), set $\lambda_{opt} = \lambda_l$ and stop; otherwise, goto next step;

Step 5 Set $\tau = \omega$;

Step 6 Compute $\theta(\tau; \lambda_l)$;

Step 7 If $\theta(\tau; \lambda_l) \leq 0$, set $\varphi_l = \tau$ and goto next step; otherwise set $\tau = \eta\tau$ and goto step 6;

Step 8 Set $\lambda_{l+1} = \lambda_l - \varphi_l h(\lambda_l)$; set $l = l + 1$ and goto Step 3.

6.3.4 Application to the drying section

The algorithm proposed has been applied to the design of fault-tolerant controller for drying section of a paper machine producing superthin condenser tissue.

The operation objective of drying section is to control the reel moisture content within quality specification. In Chapter 2, a discrete time model for the drying section using "equivalent cylinder" has been developed. The continuous model corresponding to this discrete model can be represented by

$$\dot{T}_d(t) = -0.0324T_d(t) - 8.024 \times 10^{-4}Bw_1(t) - 9.733 \times 10^{-4}Ms_1(t)$$
$$-0.01296Sp(t) + 1.491 \times 10^{-3}T_s(t) + 0.03732P(t) \qquad (6.158)$$

$$\dot{M}s(t) = -0.2510Ms(t) - 1.005T_d(t) + 0.05274Bw_1(t)$$
$$+0.0507Ms_1(t) + 0.4572Sp(t) \qquad (6.159)$$

where Ms is reel moisture content, Ms_1 the moisture content entering the drying section, Bw_1 the dry basis weight, T_d the temperature of the "equivalent cylinder", P the steam pressure, Sp the machine speed, and T_s the steam temperature. The temperature of the "equivalent cylinder" is a linear combination of that of all drying cylinders

$$T_d(t) = 0.080902T_{d1}(t) + 0.1061T_{d2}(t) + 0.1211T_{d3}(t)$$
$$+ \quad 0.1721T_{d4}(t) + 0.1625T_{d5}(t)$$
$$+ \quad 0.1781T_{d6}(t) + 0.1832T_{d7}(t) \qquad (6.160)$$

Denoting
$$x(t) = (T_d(t) \; Ms(t))^T \quad u(t) = P(t)$$
$$r(t) = (Bw_1(t) \; Ms_1(t) \; Sp(t) \; T_s(t))^T$$

as system state vector, control vector and disturbance vector, respectively, the state equation model of the drying section is described as

$$\dot{x}(t) = Ax(t) + Bu(t) + Dr(t) \qquad (6.161)$$

where
$$A = \begin{pmatrix} -0.0324 & 0.0 \\ -1.005 & -0.2510 \end{pmatrix} \quad B = \begin{pmatrix} 0.03732 \\ 0.0 \end{pmatrix}$$
$$D = \begin{pmatrix} -8.024 \times 10^{-4} & -9.733 \times 10^{-4} & -0.01296 & 1.491 \times 10^{-3} \\ 0.05274 & 0.0507 & 0.4572 & 0 \end{pmatrix}$$

The sensors in the drying section are for the measuring the surface temperatures of the seven drying cylinders, the steam pressure and temperature, and machine speed. Among them, the cylinder surface temperature sensors are most often to fail. The objective of the controller design is to improve the fault-tolerance of the control system to the surface temperature failures.

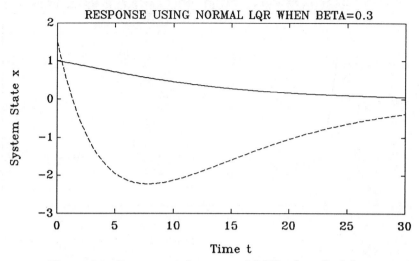

Figure 6.7: Response using normal LQR when $\beta=0.3$

Defining the performance criteria function as

$$J = \frac{1}{2} \int_0^\infty e^{\sigma t}(x^T(t)Qx(t) + u^T(t)Ru(t))dt \qquad (6.162)$$

where

$$Q = \rho Q_0 \qquad Q_0 = \begin{pmatrix} 1.0 & 0.0 \\ 0.0 & 1.0 \end{pmatrix} \qquad R = 1.0$$

The optimal values of σ and ρ which give the highest fault-tolerance are obtained using the algorithm given in Section 6.3.2 as

$$\sigma_{opt} = 0.7422, \qquad \rho_{opt} = 8.3085$$

Figures 6.7 and 6.8 indicate system responses with initial state $x(0) = (1.0\ \ 1.5)^T$ when the gain degradation factor for surface temperatures is reduced to $\beta = 0.3$. It is seen from the figures that the performance of the normal linear optimal control system becomes unsatisfactory with slow response and large overshoot. However, the linear optimal controller with high fault-tolerance design still performs well. Thus the effectiveness of the proposed algorithm is proven.

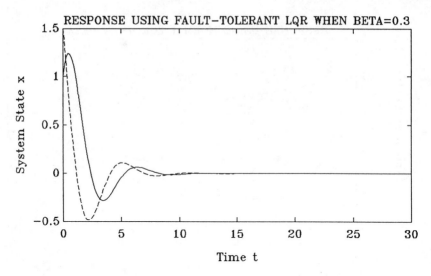

Figure 6.8: Response using fault-tolerant LQR when β=0.3

6.4   Conclusions

In this chapter, we proposed two techniques for fault-tolerant control. The first technique was applied to a pressurized paper machine headbox. This technique included fault detection schemes and state estimator reorganization schemes after the occurrence of a failure. The fault detection was accomplished by a bank of Kalman filters using the analytic redundancy among the outputs of dissimilar measurement instruments. The state estimator was reorganized to give the most accurate state estimate in normal condition and when some sensor has failed. A decoupling control law was realized using the state estimate. The technique had reasonable real-time computational burden and was easy to be implemented. The most attractive advantage of the technique was its inherent ability for automatic reorganization of system structure in the presence of a failure. The second technique was applied to the drying section of a paper machine. The fault-tolerance of linear quadratic optimal system against actuator and sensor faults was analysed. An iterative algorithm was proposed to design systems with highest fault-tolerance by selecting optimal weighting factors. Simulation results had shown good performance of the techniques. The results obtained are promising for industrial applications.

6.5 References

Anderson B.D.O and Moore J.B. Linear Optimal Control, Englewood Cliffs, NJ: Prentice-Hall, 1971

Borisson U. Self-tuning regulators for a class of multivariable systems. Automatica 1979; 15:209-215

Deckert J.C., Desai M.N., Deyst J.J. and Willsky A.S. F-8 DFBW sensor failure identification using analytic redundancy. IEEE Trans. Autom. Control 1977; AC-22:795-803

Desouze E. and Bhattacharyya S.P. Controllability, observability and the solution of AX-XB=C. Linear Algebra Appl. 1981; 3:167-188

D'Hulster F., De Keyser R.M.C. and Van Cauwenberghe A.R. Simulation of adaptive controllers for a paper machine headbox. Automatica 1983; 19:407-414

Eterno J.S., Looze D.P., Weiss J.L. and Willsky A.S. Design issues for fault-tolerant restructurable aircraft control. Proc 24th IEEE Conf. on Decision and Control, Fort Lauderdale,1985, pp 900-905

Fjeld M. Application of modern control concepts on a kraft machine. Automatica 1978; 14:107-117

Gilbert E.G. The decoupling of multivariable system by state feedback. SIAM J. Control 1969; 7:50-63

Gunn D.J. and Sinha A.K. Nonlinear control of a paper-stock flowbox. Proc. IEE, 1983; 130D:131-136

Keel L.H., Bhattacharyya S.P. and Howze J.W. Robust control with structured perturbations. IEEE Trans. Automat. Contr. 1988; AC-29:68-77

Lebeau B., Arrese R., Bauduin S., Grobet R. and Foulard C. Noninteracting multivariable paper machine headbox control: some comparisons with classical loops. Proc 4th IFAC Conf. on Instr. and Autom. in the Paper, Rubber, Plastics and Polym. Industr., Gent, 1980, pp 227-238

Locatelli A., Scattolini R., and Schiavoni N. On the design of reliable robust decentralized regulators. Large Scale System 1986; 10:95-113

Looze D.P., Krolowski S.M., Weiss J.I., Eterno J.S. and Gully S.W. An approach to restructurable control system design. Proc 23rd IEEE Conf. on Decision and Control, Las Vegas, 1984, pp 1392-1397

Mariton M. and Bertrand P. Reliable flight control systems: components placement and feedback synthesis. Proc 10th IFAC World Congress, Munich, 1987, pp 150-154

Patel R.V., Toda M. and Sridhar B. Robustness of linear quadratic state feedback design in the presence of system uncertainty. IEEE Trans. Autom. Contr. 1977; AC-22:945-949

Petersen I.R. A procedure for simultaneously stabilizing a collection of single input linear systems using non-linear state feedback control. Automatica 1987; 23:33-40

Polak E. Computational Methods in Optimization: A Unified Approach. Academic Press, 1971

Shimenura E. and Fujita M. A design method for linear state feedback systems processing integrity based on a solution of Riccati-type equation. Int. J. Control 1985; 42:887-899

Siljak D.D. Reliable control using multiple control systems. Int. J. Control 1980; 33:303-329

Vidyasagaar M. and Viswanadhm N. Reliable stabilization using multi-controller configuration. Automatica 1985; 21:599-602

Willsky A.S. A survey of design methods for failure detection in dynamic systems. Automatica 1976; 12(1):601-611

Willsky A.S. Detection of abrupt changes in dynamic systems. Lecture Notes in Control and Information Sciences, (Ed. M. Basseville and A. Benveniste), Vol 77. Springer-Verlag, 1986

Xia Q., Sun Y. and Zhou C. An optimal decoupling controller with application to the control of paper machine headboxs. Presented at the 4th IFAC Symposium on CAD in Control Systems, Beijing, P.R. China, 1988

Xia Q., Rao M., Shen X. and Ying Y. Systematic modeling and decoupling design of a pressurized headbox. Proc Canadian conference on electrical and computer engineering, Vancouver, 1993, pp 962-965

Xia Q. and Rao M. Fault-tolerant control of a paper machine headbox. Journal of Process Control 1992; 3:271-278

Yedavalli R.K. Improved measures of stability robustness for linear state space model. IEEE Trans. Autom. Contr. 1985; AC-30:577-579

CHAPTER 7
FUZZY CONTROL

In control engineering, the conventional method of designing controllers starts with a precise mathematical model of the process. The desired behaviours of the closed-loop system are also expressed mathematically. All these mathematical information is combined and the equations can usually be solved to give a mathematical expression of control to be carried out. The sophisticated automation application from aircrafts and spacecrafts to chemical processes is a tribute to the success of this method and its numerous variations. However, this success can mainly be achieved when the models are given as linear equations.

Many complex processes, unfortunately, do not yield to this design procedure, since accurate mathematical models are not available or too complicated to design controller. As such processes are already controlled by the skilled human operators it follows that this is an area where fuzzy control could be useful.

In this chapter, we will discuss the fuzzy optimal control (Section 7.1) and fuzzy-precise combined control (section 7.2) as well as their applications to the paper machines. Figure 7.1 shows the schematic diagram of the fuzzy control system.

7.1 Fuzzy Optimal Control

7.1.1 Introduction

Since fuzzy controller was first successfully applied to a steam engine process control by Mamdani in 1974 (Mamdani, 1974), it has brought up wide interests. Many results have been reported both in theory and application.

193

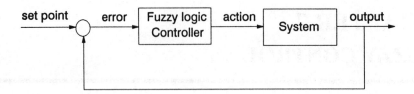

Figure 7.1: Schematic diagram of fuzzy control system

Fuzzy controller has been recognized as a useful tool in process control (Tong, 1977, 1983; Gupta, 1981; Sugeno, 1985).

However, there are still some difficult problems in the synthesis of fuzzy controllers. The main problem, as pointed by Czogala and Pedrycz (1982) is that "the general scheme of their designing and optimization is not available". The usual procedure of fuzzy controller design can be divided into two steps. Firstly, fuzzy control rules are induced from the experience of process operators. They are linguistic conditional statements relating the input variables of controller to its output variables, such as "IF PE (pressure error) is NS (negative small) and CPE (change in pressure error) is NS (negative small), THEN HC (heat input change) is PM (positive medium)". Secondly, some fuzzy subsets are defined in the universe of discourse of both input and output variables or the controller which assigns every quantized variable, either input or output, a membership value. By carrying out a series of fuzzy logic operations under these fuzzy subset definitions and control rules obtained from the first step, a decision table, or called as look up table, is produced. This table relates the quantized input and output variables of fuzzy controller with a simple mapping relation which can be implemented easily into a digital computer.

Of these two steps, the first one is more important because fuzzy control rules will actually determine the behavior of the fuzzy controller (Tong, 1976). But unfortunately, there is no general principle given in this step to guide the induction of fuzzy control rules. Designers can only depend on their own skill in collecting the experiences of operators and abstracting them into fuzzy control rules. Different designers may give different forms of the induced rules and thus results in different performance of the fuzzy controller. On the other hand, if process operators themself could not maintain an optimal operating condition, the control rules induced from their experience will not be optimal either. Therefore, the design of a fuzzy controller is some kind of art rather than a technique, hence the application is limited. To make fuzzy controller more practical, it is necessary to develop a systematic design procedure.

Here we present a systematic synthesis method which gives a general scheme for designing fuzzy controller. The resulted controller is optimal with respect to the performance index $J = \sum_{k=1}^{T_f} (aN_{A(k)} + bN_{B(k)})$.

In the following section, a briefly review of the fuzzy logic theory as well as some preliminary definitions are given. Section 7.1.3 presents the systematic synthesis method. In Section 7.1.4, the application of this design method to the moisture content control of a paper machine is presented.

7.1.2 Fundamental of fuzzy logic and fuzzy optimal control

The basic theory of fuzzy logic was developed by Zadeh (1975). Here, only a few notations and definitions to be used in the systematic design method of fuzzy optimal controller will be reviewed.

A fuzzy subset A is defined as

$$A = \mu(u)/u; u \in U$$

where

$\mu(u) \in [0, 1]$ is the membership value of the subset A at the point u;

U is the universe of discourse of the variable whose linguistic label is A.

In fuzzy set theory, adjectives or hedges (such as *very, more or less, not*) are defined as mathematical functions of $\mu(u)$. Thus,

$$\mu_{veryA}(u) = \mu_A^2(u); u \in U$$

and

$$\mu_{notA}(u) = 1 - \mu_A(u); u \in U$$

In addition to the above hedges, there are logical connectives OR, AND defined as:

$$\mu_{AORB}(u) = \mu_A(u)V\mu_B(u); u \in U$$

$$\mu_{AANDB}(u, v) = \mu_A(u)\Lambda\mu_B(v); u \in U, v \in V$$

The concept of fuzzy optimal control was first proposed by Bellman and Zadah(1970) when they studied the multistate decision making problems (Bellman and Zadeh, 1970), and was further studied by many other researchers (Fung and Fu, 1977; Dubois and Prada, 1980; Kacprzyk, 1982, 1983; Czogala and Pedrycz, 1982). Multistate decision making is fuzzy sets application in control theory (Tong, 1983), in which the idea of optimization is to make the

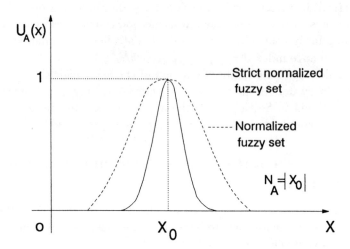

Figure 7.2: Definition of strict normalized fuzzy set and its norm

states and controls of a system as close to the desired states and controls as possible. This means that the tracks of optimal (desired) states and optimal (desired) controls must be known before the optimal control algorithm can be designed. For a process control system, this is difficult, therefore such kind of fuzzy optimal control is more suitable for management system rather than for process control (Li, et al., 1985).

The fuzzy controller to be proposed here is optimal in the different sense to the one mentioned above. To distinguish these two kinds of fuzzy controls, the one we designed is called optimal fuzzy controller.

Before presenting the optimal fuzzy controller, two preliminary definitions should be introduced.

Definition 1 Let X be a universe of discourse, x a generic element of X, and A a fuzzy subset of X with its membership function $\mu_A(x) \in [0, 1]$. A is called strict normalized iff $\exists x_0 \in X$, $\mu_A(x_0) = 1$, and $\forall x \in X$, $x \neq x_0$, $\mu_A(x) < \mu_A(x_0)$.

Definition 2 For a strict normalized fuzzy subset A, its norm N_A exists, and $N_A = \{| X_0 | - \mu_A(x_0) = 1\}$.

These two definitions can be illustrated with Figure 7.2.

7.1.3 Systematic synthesis procedure of optimal fuzzy controller

For simplicity, a single-input single-output system is considered.

To build a fuzzy linguistic model of the controlled process, the fuzzy identification method (Li and Liu, 1980), a tool for establishing a fuzzy linguistic rule formed model from input-output data, is modified to fulfil the task.

Referential fuzzy subsets A_1, A_2,..., A_{n1} and B_1, B_2, ..., B_{n2} are defined on the universe of discourse U and Y of input and output of the process. The numbers of fuzzy subsets, n_1 and n_2, are assigned by designers according to the control precision requirement. All subsets should be strict normalized and satisfy the condition that $\forall u \in U, \exists \mu_{Ai}(u) \geq C_1$ and $\forall y \in Y, \exists \mu_{Bj}(y) \geq C_2$, where C_1 and C_2 are real numbers lie in the interval [0,1]. Within these referential fuzzy subsets, all measured values of input and output of the process will belong to certain fuzzy subset with a membership function μ.

Because fuzzy controllers are usually implemented with digital computers, only discrete input and output sampling data are available. In process control system, the sampling interval should be much smaller than the response time of process. In reference (Li and Liu, 1980), the rules of fuzzy model take the following form

$$if \ u(k-l) = A_i \ and \ y(k-m) = B_j \ then \ y(k) = B_h \qquad (7.1)$$

$$(i = 1, ..., n_1, j = 1, ..., n_2, h \in \{1, ..., n_2\})$$

where k, l and m are sample times.

For optimization, the rules of fuzzy model are modified as

$$if \ U = A_i \ and \ Y = B_j \ then \ after \ T = k_{ij}, \ Y_f = B_h \qquad (7.2)$$

$$(i = 1, ..., n_1, j = 1, ..., n_2, h \in \{1, ..., n_2\})$$

where k_{ij} is the number of time interval, and T is a nonfuzzy time variable. For simplicity, we denote this rule as the (i×j)th rule. In doing so, it should be noted that the fuzzy model is more static and less dynamic, it take the advantages of both static and dynamic models.

With this modification, the identification procedure will also become simpler and more direct. Collecting the sampling data of input u and output y of the process during regular operation or through a series of tests and clustering them according to maximum membership function principle, a group of rules in the form of expression (7.2) can be obtained. The maximum number of such rules should be $n_1 \times n_2$ and to this point the fuzzy model is completed. But it is not always necessary to have the maximum number of rules, the reason will be discussed later.

After the fuzzy model is obtained, the synthesis of an optimal fuzzy controller can be performed.

The first step is to calculate the fuzzy cost for every rule in fuzzy model, i.e.,

$$J_{ij} = aN_{Ai} + bN_{Bj} \quad (i = 1, ..., n_1, j = 1, ..., n_2) \tag{7.3}$$

where J_{ij} is the cost of the (i×j)th rule, a and b are weighting factors which are selected somewhat like the weighting matrices Q and R in quadratic optimal problems, and N_{Ai} and N_{Bj} are norms of fuzzy subsets A_i and B_j, respectively.

With these cost value J_{ij}, a group of new rules in fuzzy model is obtained

$$if \ U = A_i \ and \ Y = B_j \ then \ Y_f = B_h \ with \ J_{ij} \tag{7.4}$$

$$(i = 1, ..., n_1, j = 1, ..., n_2, h \in \{1, ..., n_2\})$$

With these new rules, the optimal fuzzy control rules can be obtained by using dynamic programming algorithms which minimize the fuzzy performance index

$$J = \sum_{k=1}^{T_f} (aN_{A(k)} + bN_{B(k)}) \tag{7.5}$$

where k is the sample time, T_f is the time for the response of the system, $A(k) \in \{A_1, A_2, ..., A_{n_1}\}$ and $B(k) \in \{B_1, B_2, ..., B_{n_2}\}$ are fuzzy subsets of the input u and output y of the process at time interval k, respectively, $N_{A(k)}$ and $N_{B(k)}$ are norms of $A(k)$ and $B(k)$.

The second step is optimization by using dynamic programming method. Introducing a target fuzzy subset B_0 of the universe of discourse Y, $B_0 \in \{B_1, B_2, ..., B_{n_2}\}$, the output y should be controlled to go inside the fuzzy subset B_0. When all rules including B_0 in their conclusions are found, record the fuzzy cost $J_{ij}^{(1)}$ of each rule, and denote the fuzzy subset of y in its premise as $B_{i0}^{(1)}$. From $B_{i0}^{(1)}$, using the same heuristic search approach, a group of fuzzy subsets $B_{ji}^{(2)}$ can be formulated with the corresponding fuzzy cost $J_{ij}^{(2)}$. Repeating the same search process, a tree structure can be obtained, as shown in Figure 7.3. Using dynamic programming in accordance with this tree structure, the minimum cost line can be found for every B_j (except B_0) to reach B_0. Thus, a series of optimal fuzzy control rules, in the form: if $y = B_j$ then $u = A_i$, are obtained which will make the performance index (7.5) minimum.

For the regulator problem, usually the setpoint is fixed, hence only one B_0 is selected and thus only one series of optimal search operation is needed to get one group of optimal fuzzy control rules. If servo problem is considered,

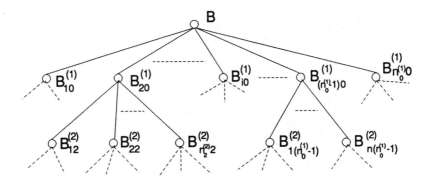

Figure 7.3: Tree structure of heuristic searching

several series of optimal search operations are needed to find optimal fuzzy control rules.

After optimal fuzzy control rules are obtained, the rest of the design procedure is essentially the same as that of the ordinary fuzzy controllers, which can be found in many textbooks about fuzzy sets (Zadeh, 1975), and we will not discuss in detail.

7.1.4   Application to moisture control

The proposed optimal fuzzy controller has been implemented in a computer control system for a paper machine.

Basis weight and moisture content are two important quality variables of paper product. Moisture content is controlled by steam pressure which is changed through the steam valve opening. Because there is a PID controller for steam pressure control, only the relation between steam pressure and moisture content is considered in fuzzy control, i.e., taking steam pressure as input u and moisture content as output y.

Two fuzzy variables are defined:

Δu — steam pressure deviation, i.e., the difference between the measured and desired values of u.

Δy — moisture content deviation, the difference between the measured and desired values of y.

From the operating conditions of the paper machine, the universe of discourse of ΔU and ΔY are defined as [-60,60] and [-15,15] , respectively. The number of referential fuzzy subsets n_1 and n_2 are determined according to the required control precision. The higher the control precision required, the

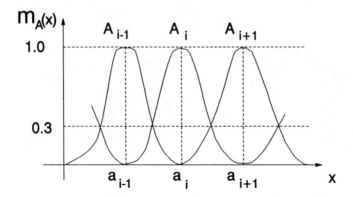

Figure 7.4: Fuzzy subsets with the required membership function

larger the number of referential fuzzy sets needed. Certain skill and experience are needed to determine n_1 and n_2. Taking compromise between the control precision and computational simplicity, thirteen referential subsets are defined on each of these two universes of discourse, with the membership functions defined as

$$A_i(\delta u) = e^{\frac{-(\delta u - \alpha_i)^2}{\gamma_1}} \qquad (i = 1, ..., 13) \qquad (7.6)$$

$$B_j(\delta y) = e^{\frac{-(\delta y - \beta_j)^2}{\gamma_2}} \qquad (j = 1, ..., 13) \qquad (7.7)$$

where α_i is the central value of the subsets A_i and β_j is that of B_j, γ_1 and γ_2 are parameters which are determined to give the membership value of 0.3 on the intersections of two fuzzy subsets. Figure 7.4. shows the shape of the defined fuzzy subsets. With these fuzzy subsets, the identification of fuzzy model is then carried out. As mentioned early, the minimum number of rules in fuzzy model could be $n_1 \times n_2^2$, i.e., 13^3. However, not all rules are necessary in fuzzy controller design. The aim of the control is to maintain the moisture content at its setpoint, i.e., to keep δy equal to zero, and thus in the fuzzy controller the output error δy is controlled within fuzzy subset B_7. Therefore in the identification, only those rules which express the process response in the direction from any fuzzy subset towards B_7 are recorded. Finally only 143 rules of the fuzzy model in the form of expression (7.2) are registered through identification, and they represent the process under the normal condition.

Based on these rules, the optimization design is performed. Weighting factors a and b are selected in order to get several different groups of optimal fuzzy control rules. By comparison among these groups, one of them, which emphasizes the stability of δu instead of δy, is chosen for application to the practical control. This group of control rules is designed with $a=0.7$ and $b=0.3$, so that the control action is not too strong. It is expressed in Table 1.

Table 1. Optimal fuzzy control rules with a=0.7, b=0.3.

If y=	B_1	B_2	B_3	B_4	B_5	B_6	B_7	B_8	B_9	B_{10}	B_{11}	B_{12}	B_{13}
Then u=	A_1	A_2	A_3	A_4	A_5	A_6	A_7	A_8	A_9	A_{10}	A_{11}	A_{12}	A_{13}

These rules are transformed to form a look up table by ordinary fuzzy control design procedure and then implemented into a computer. The control results are presented in Figure 7.5. Figure 7.6 shows the control results of deterministic optimal control system. From the comparison of these results, it can be seen that the optimal fuzzy controller performs as well as the deterministic optimal control.

7.2 Fuzzy-Precise Combined Control

7.2.1 Introduction

The shortcomings of precise control, such as its dependence on the accuracy of process model can be avoided by applying fuzzy control. However, fuzzy control also has its own weakness, such as poor control precision and complexity when it is applied to multi-input multi-output (MIMO) systems. In industrial applications, it is always the objected to design a simplest control system with highest control precision. For this reason, it is natural to consider the combined advantages of fuzzy and precise controls.

In this section, a fuzzy-precise combined (FPC) control approach is proposed to make the design of multi-input multi-output control systems simple and efficient. This approach keeps the advantages of both fuzzy and precise control system and avoid their shortcomings.

In Section 7.2.2, the appropriate control principle as the basic idea of the fuzzy-precise combined control approach is introduced. This new approach consists of three parts: (1) a fuzzy criterion which determines the attributions of subsystems of FPC system; (2) an FPC model of the system; and (3) an FPC control strategy. The application of FPC control to paper machines is presented in Section 7.2.3.

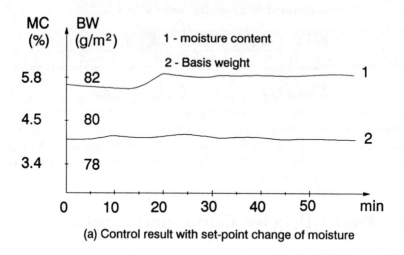

(a) Control result with set-point change of moisture

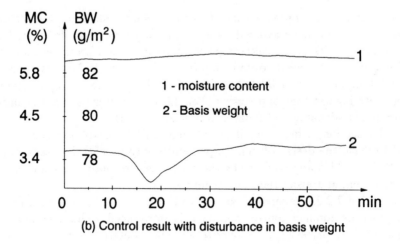

(b) Control result with disturbance in basis weight

Figure 7.5: Results of fuzzy optimal control

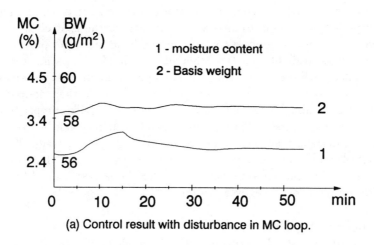

(a) Control result with disturbance in MC loop.

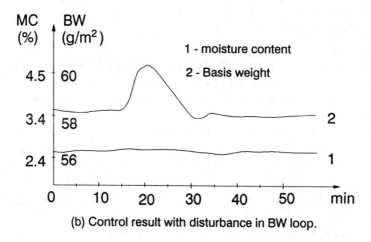

(b) Control result with disturbance in BW loop.

Figure 7.6: Results of deterministic optimal control

7.2.2   FPC control

The appropriate control principle

A control system is usually expected to have high precision and good performance. Advanced but complex control algorithms are often employed for this purpose, which requires an accurate models of the controlled processes. However, in industrial process control, on one hand, accurate process models are not always available; on the other hand, not all the parameters should be controlled with high precision. For example, some process variables other than the final quality variables allow certain drifts from their desired values. Considering the above two issues, a new concept namely "appropriate control principle" is introduced. "Appropriate control principle" can be expressed as follows: In designing an industrial process control system, trade-off between the control benefit and design effort should be taken, such that the control objective can be appropriately determined for the largest control benefit and the least design effort whenever possible.

FPC and fuzzy criterion

An MIMO industrial process usually can be physically divided into several interactive subprocesses. In other words, a control system can be decomposed into several subsystems. For some subsystems, it is easy to build accurate mathematical models, but for others, the accurate models may be difficult or even impossible to build.

Such MIMO systems are not appropriate to apply unitary precise control approach or fuzzy control approach. Using precise control only will make the control system design very difficult and fuzzy control alone will lose some attainable precisions.

Following the appropriate control principle, a fuzzy-precise combined (FPC) control approach is proposed. Precise control is used for subsystems whose mathematical models are easy to build. For subsystems whose mathematical models are not amenable, the fuzzy control is used. Therefore the combined approach will be able to keep high control precision and avoid difficulties in system design.

To apply fuzzy-precise combined control approach, the first thing is to determine the attribution of all subsystems, that is, to determine the subsystem is fuzzy or precise. This can be done by using experience or criteria. Here a fuzzy criterion is suggested.

Consider the degree of difficulty in building an accurate mathematical model for a subsystem as the measure of design effort, and the degree of requirement to the control precision of a subsystem as the measure of control benefit. The fuzzy criterion can be expressed as follows.

Suppose an MIMO system P_c is physically divided into several subsystems. Whether a subsystem should be fuzzy or precise can be determined by the fuzzy criterion

$$\underline{F} = (\underline{D} * \underline{R_1}) \cup (\underline{E} * \underline{R_2}) \qquad (7.8)$$

where, $\underline{F}$ – fuzzy determination set, which includes two subsets, i.e., $\underline{F} = \{\text{precise approach, fuzzy approach}\} = \{\underline{F_1}, \underline{F_2}\}$;

$\underline{D}$ – fuzzy factor set 1, which expresses the degree of difficulty in building an accurate mathematical model, includes four subsets, i.e., $\underline{D} = \{\text{not difficult, a little difficult, difficult, very difficult}\} = \{\underline{D_1}, \underline{D_2}, \underline{D_3}, \underline{D_4}\}$;

$\underline{E}$ – fuzzy factor set 2, which expresses the degree of requirement to the control precision and includes four subsets, i.e., $\underline{E} = \{\text{low, mid, high, very high}\} = \{\underline{E_1}, \underline{E_2}, \underline{E_3}, \underline{E_4}\}$;

$\underline{R_1}, \underline{R_2}$ – fuzzy transfer matrices.

For the ith subsystem P_i, equation(7.8) means that

$$f_i = (d_i * \underline{R_1}) \cup (e_i * \underline{R_2}) \qquad (7.9)$$

where

$f_i = (\mu_{\underline{F_1}}(P_i), \mu_{\underline{F_2}}(P_i))$;

$d_i = (\mu_{\underline{D_1}}(P_i), \mu_{\underline{D_2}}(P_i), \mu_{\underline{D_3}}(P_i), \mu_{\underline{D_4}}(P_i))$;

$e_i = (\mu_{\underline{E_1}}(P_i), \mu_{\underline{E_2}}(P_i), \mu_{\underline{E_3}}(P_i), \mu_{\underline{E_4}}(P_i))$.

The attribution of the ith subsystem is determined by the comparison between $\mu_{\underline{F_1}}(P_i)$ and $\mu_{\underline{F_2}}(P_i)$, i.e., select $\underline{F_i}$ such that

$$\mu_{\underline{F_i}}(P_i) = max(\mu_{\underline{F_1}}(P_i), \mu_{\underline{F_2}}(P_i)) \qquad (7.10)$$

Using this fuzzy criterion, attribution of every subsystem can be determined, and then the fuzzy-precise combined system P_c is obtained, which maximizes the index

$$J_{\underline{F}} = \sum_{j}(\mu_{\underline{F_i}}(P_j)), \quad \underline{F_i} \in \{\underline{F_1}, \underline{F_2}\} \qquad (7.11)$$

In this fuzzy criterion, the fuzzy transfer matrices $\underline{R_1}$ and $\underline{R_2}$ need to be determined by process control experts.

FPC Model

After the attributions of all subsystems are determined, the next step is to build fuzzy–precise combined models.

Suppose that there are N interactive subsystems, among which N_p are precise and N_f are fuzzy, $(N_p + N_f = N)$.

For the N_p precise subsystems, their accurate mathematical models can be expressed by the discrete state space form

$$X_{pi}(k+1) \;=\; A_{pi}X_{pi}(k) + B_{pi}U_{pi}(k) + \sum_{j=1,j\neq i}^{N_p} L_{pij}X_{pj}(k)$$

$$+ \sum_{h=1,h\neq i}^{N_f} L_{fih}Y_{fh}(k) \qquad i = 1, ..., N_p \qquad (7.12)$$

$$Y_{pi}(k) = C_{pi}X_{pi}(k) \qquad i = 1, ..., N_p \qquad (7.13)$$

where X_{pi} and U_{pi} are $n_i \times 1$ states and $m_i \times 1$ control vectors of the ith precise subsystem respectively, A_{pi} and B_{pi} are $n_i \times n_i$, $n_i \times m_i$ parameter matrices, L_{pij} and L_{fih} are $n_i \times m_j$ and $n_i \times 1$ interacting matrices, Y_{fh} is the output of the hth fuzzy subsystem. Suppose that there are a total of n_p states in the N_p precise subsystems, then $n_p = \sum_{i=1}^{N_p} n_i$, and these states can be noted as X_1, ..., X_{n_p}.

The accurate mathematical models of precise subsystems can also be expressed in the form of transfer functions, in accordance with the need of the control strategy.

For the N_f fuzzy subsystems, the fuzzy control rules can be directly obtained from the experience of operators. Section 7.1 has given a systematic method for designing optimal fuzzy controller, in which the fuzzy linguistic model takes the following form

$$\begin{cases} IF \;\; U_{fj} = \underline{A}^j_{i_{u_{fj}}} \;\; and \;\; Y_{fj} = \underline{B}^j_{i_{y_{fi}}} \\ \quad and \;\; interactions \;\; in \;\; X_1, .., X_{n_p}, Y_{f_1}, .., Y_{f_{N_f}} \\ \\ THEN \;\; after \;\; T = k_{(i_{u_{fj}}, i_{y_{fj}}, g_{y_{fj}})}, Y_{fj} = \underline{B}_{g_{y_{fj}}} \\ \quad j = 1, ..., N_f \end{cases} \qquad (7.14)$$

where $i_{u_{fj}} = 1, ..., s_{u_{fj}}, i_{y_{fj}} = 1, ..., s_{y_{fj}}, g_{y_{fj}} \in \{1, 2, ..., s_{y_{fj}}\}$; $\underline{A}^j_{i_{u_{fj}}} \; (j=1, ..., N_f)$ are fuzzy subsets defined by the universe of discourse $U_{fj} \; (j = 1,..., N_f)$; $\underline{B}^j_{i_{y_{fj}}} \; (j=1, ..., N_f)$ are fuzzy subsets defined by the universe of discourse Y_{fj} $(j = 1,..., N_f)$; $S_{u_{fj}}$ and $S_{u_{fj}} \; (j = 1, ..., N_f)$ are the numbers of fuzzy subsets $\underline{A}^j_{i_{u_{fj}}}$ and $\underline{B}^j_{i_{y_{fj}}}$ respectively; and $k_{(i_{u_{fj}}, i_{y_{fj}}, g_{y_{fj}})}$ is the sampling time.

The above fuzzy linguistic model is the fuzzy model of the jth fuzzy subsystem. The appeared complexity of this model is due to the consideration of time factor T and interactions with other subsystems. In practice, only significant interactions are considered so that the model can be much simpler.

FPC Algorithm

Knowing the combined system model (7.12), (7.13) and (7.14), the fuzzy-precise combined optimal control algorithm can be developed.

The total cost function is

$$J_c = J_p + J_f \tag{7.15}$$

where

$$J_p = \sum_{i=1}^{N_p} \{ \sum_{k=0}^{T_f-1} (\frac{1}{2} \parallel X_{pi}(k) \parallel_{Qi}^2 + \frac{1}{2} \parallel U_{pi}(k) \parallel_{Ri}^2) \} = \sum_{i=1}^{N_p} J_{pi} \tag{7.16}$$

$$J_f = \sum_{j=1}^{N_f} \{ \sum_{k=1}^{T_f} [aN_{A_{f u_j}(k)} + bN_{\underline{B}_{f y_j}(k)}] \} = \sum_{j=1}^{N_f} J_{fj} \tag{7.17}$$

It is clear from equations(7.16) and (7.17) that J_p is the precise cost function and J_f is the fuzzy cost function. Both of them are the sum of the subsystem cost functions. Therefore, the total cost function J is decomposed into N sub-cost functions and the global optimal control problem of fuzzy-precise combined system is transformed into N subsystem optimal control problems.

For the ith precise subsystem the optimal algorithm is given by the following recursive formula

$$K_i(k) = Q_i + A_{pi}^T K_i(k-1)[I + B_{pi} R_i^{-1} B_{pi}^T K_i(k-1)]^{-1} A_{pi} \tag{7.18}$$

$$S_i(k) = A_{pi}^T \{ K_i(k-1)[I + B_{pi} R_i^{-1} B_{pi}^T K_i(k-1)]^{-1}[-B_{pi} R_i^{-1} B_{pi}^T S_i(k-1)$$

$$+ \sum_{j=1}^{N_p} L_{pij} X_{pj}(k) + \sum_{h=1}^{N_f} L_{fih} Y_{fh}(k)] + S_i(k-1) \} \tag{7.19}$$

$$U_{pi}(k) = -R_i^{-1} B_{pi}^T A_{pi}^{-T} \{ [K_i(k) - Q_i] X_{pi}(k) + S_i(k) \} \tag{7.20}$$

$$X_{pi}(k+1) = A_{pi} X_{pi}(k) + B_{pi} U_{pi}$$

$$+ \sum_{j=1}^{N_p} L_{pij} X_{pj}(k) + \sum_{h=1}^{N_f} L_{fih} Y_{fh}(k) \tag{7.21}$$

The control law (7.20) will minimizes the cost function

$$J_{pi} = \sum_{0}^{T_f-1} (\frac{1}{2} \parallel X_{pi}(k) \parallel_{Q_i}^2 + \frac{1}{2} \parallel U_{pi}(k) \parallel_{R_i}^2) \tag{7.22}$$

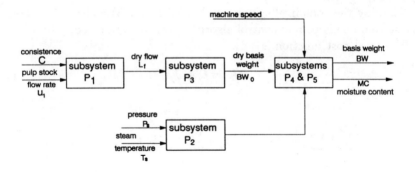

Figure 7.7: Subsystems of condenser paper machine

Using the optimal fuzzy controller design procedure based on the fuzzy model (7.14), the following optimal fuzzy control rule can be obtained:

IF $Y_{fj} = \underline{B}^j_{i_{y_{fj}}}$ and interactions in $X_1, ..., X_{n_p}, Y_{f1}, ..., Y_{fn_f}$
THEN $U_{fj} = \underline{A}^j_{i_{u_{fj}}}$ $j = 1, ..., N_f$

which minimizes

$$J_{fj} = \sum_{k=1}^{T_f} (a N_{\underline{A}_{fuj}(k)} + b N_{\underline{B}_{fyi}(k)}) \tag{7.23}$$

7.2.3   Application of FPC system to paper machine

The FPC system is implemented in a paper machine which produces condenser paper. The whole paper machine system can be decomposed into five subsystems, as shown in Figure 7.7. For subsystems P_1 and P_2, the accurate mathematical models can be built using mechanism analysis without uncertainties, thus P_1 and P_2 are precise subsystems. For subsystems P_3, P_4 and P_5, it is difficult to obtain accurate models so that they are treated as fuzzy subsystems.

Subsystem P_1 is considered as an ideal system without time delay. By mechanism analysis, its model is

$$L_f(k) = k_1 u_1(k) = k_2 c(k) \tag{7.24}$$

where k_2 is a function of speed V.

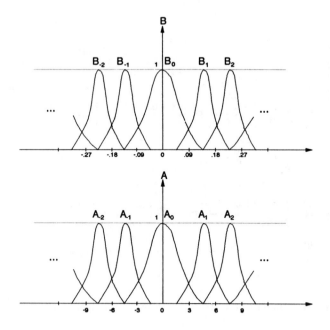

Figure 7.8: Fuzzy subsets definition of subsystem P_3

The model of subsystem P_2 is developed from on-site test and mechanism analysis

$$T_d(s) = \frac{k_3}{(T_1 s + 1)(T_2 s + 1)} P_s(s) - \frac{k_4}{T_3 s + 1} T_s(s) \tag{7.25}$$

Based on mechanism analysis and technical data, the fuzzy model for subsystem P_3 is

$$If \quad L_f(k) = \underline{A}_i \quad Then \quad Bw_0(k) = \underline{B}_i \quad i = 0, \pm 1, \pm 2, \dots \tag{7.26}$$

where $\underline{A}_i$ and $\underline{B}_i$ are fuzzy numbers as shown in Figure 7.8.

Subsystems P_4 and P_5 are considered to include the time delay and capacities. Their fuzzy models are in the recursive forms:

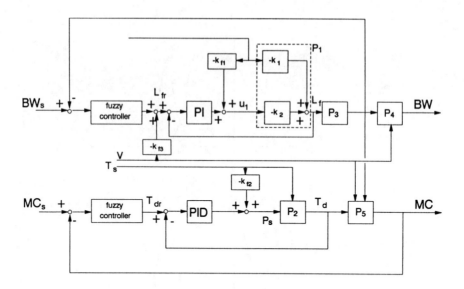

Figure 7.9: FPC control system for condenser paper machine

Subsystem P_4:

$$\begin{cases} Bw(k) = x_1(k-1) - a_1 T_d(k) \\ x_1(k-1) = x_1(k-2) - a_2 T_d(k) \\ x_1(k-2) = x_1(k-3) - a_3 T_d(k) + Bw_s(\frac{V_s}{V(k)} - 1) \\ x_1(k-i) = a_4 x_1(k-i) & i = 3,...,8 \\ x_1(k-i) = a_4 x_1(k-i) + a_5 x_1(k-i-1) & i = 9,...,14 \\ x_1(k-15) = Bw_0(k) \end{cases} \quad (7.27)$$

where Bw_s and V_s are setpoints of basis weight and speed, a_1-a_5 are constants.
Subsystem P_5:

$$\begin{cases} Ms(k) = x_2(k-1) - b_1 T_d(k) \\ x_2(k-1) = x_2(k-2) - b_2 T_d(k) \\ x_2(k-2) = b_3 Bw_0(k) - b_4 T_d(k) \end{cases} \quad (7.28)$$

where $b_1 - b_4$ are constants; Ms is reel moisture content.

Based on the above fuzzy-precise combined model, a fuzzy-precise combined control system is designed as shown in Figure 7.9.

The on-line performance of FPC control system is shown in Figure 7.10, which indicates that the system works well. Figure 7.11 shows the comparison

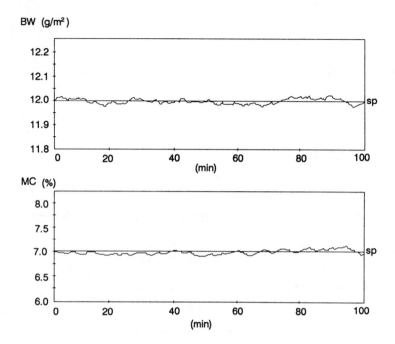

Figure 7.10: On-line control results of FPC system

of laboratory test results and FPC model estimations which indicates that the FPC model estimation well conforms to the actual values.

7.3 Conclusions

A systematic synthesis method of optimal fuzzy controller was proposed in Section 7.1. The main shortcomings of ordinary fuzzy controller, i.e., lack of systematic design and optimizing procedure, were overcome, hence the fuzzy controller becomes more practical. This new synthesis method can be carried out by a package of computer program, and an optimal fuzzy controller can be designed with a little operation process experience and system design skill. The application of this method to the control of a paper machine gave a satisfactory result.

Although what has been presented here is a simple application of the new design procedure and optimal fuzzy controller, complex applications can be developed without much difficulty, such as for multi-input and multi-output systems and heuristic optimal fuzzy control rules.

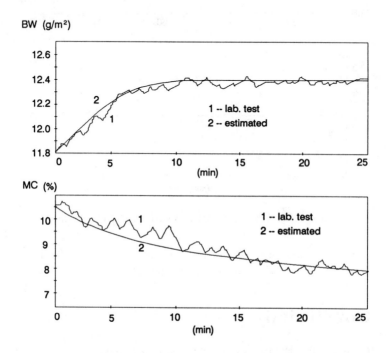

Figure 7.11: Comparison of lab. test results and FPC model estimations

There are still some shortcomings in this method such as the need of large computer memory when dealing with a multivariable system and the lack of proficiency for dynamic programming when the number of referential fuzzy subsets becomes too large. However, as a new design method of fuzzy controller, its advantages are clear and it can be applied widely.

The FPC control system was also discussed in this chapter. The FPC control approach took all advantages from both precise and fuzzy control systems. The real industrial control results showed that the fuzzy-precise combined control approach is a good solution to the control problems when the accurate process mathematical model was very difficult to obtain. The basic idea of this approach was the appropriate control principle, and its main point was to get the maximum control benefit with the minimum design and implementation efforts.

7.4 References

Åström K.J. Computer control of a paper machine— an application of linear stochastic control theory. IBM Journal 1967; 7:389-405

Bellman R.E. and Zadeh L.A. Decision making in a fuzzy environment. Manage. Sci. 1970; 17:151-169

Cegrell T. and Hedqvist T. Successful adaptive control of paper machine. Automatica 1975; 11:53-59

Czogala E. and Pedrycz W. Fuzzy rules generation for fuzzy control. Cybernetics and Systems 1982; 13:275-290

Dubois D. and Prada H. Fuzzy sets and system: Theory and Applications. Academic Press, 1980

Fung L.W. and Fu X.S. Characterization of a class of fuzzy optimal control problems. In: Gupta M.M., Saridis G.N. and Gaines B.R. Ed., Fuzzy Automata and Decision processes, North-Holland, 1977, pp 209-219.

Gupta M.M. Feedback control applications of fuzzy set theory: a survey. Proc of 8th IFAC world Congress, Kyoto, Japan, 1981, pp.761-766

Kacprzyk J. Multistage decision processes in a fuzzy environment: a survey.

In: Fuzzy Information and Decision Processes, Gupta M.M. and Sanchez E. Ed., North-Holland, 1982, pp 251-263.

Kacprzyk J. Towards 'human-consistent' multistage decision making and control models using fuzzy sets and fuzzy logic. Fuzzy Sets and Systems 1986; 10:291-298

Li B.S. and Liu Z.J. Model identification using fuzzy sets theory. Information and Control (Chinese) 1980; 3:32-38

Li P., Zhou C., Sun Y. and Ying Y. A multi-purpose robust controller and its application to paper machine control system. Proc of Int. Conf. on Industrial Process Modelling and Control, Hangzhou, China, 1985, pp 178-185

Mamdani E.H. and Aasilian S. An experiment in linguistic synthesis with a fuzzy logic controller. Int. J. Man-machine Stud. 1974; 7:1-13

Mamdani E.H. Application of fuzzy algorithms for control of simple dynamic plant. Proc. IEE 1974; 121D:1585-1588

Sugeno M. An introductory survey of fuzzy control. Information Sciences 1985; 36:59-83

Tong R.M. A control engineering review of fuzzy systems. Automatica 1977; 13:559-569

Tong R.M. Fuzzy control systems: A retrospective. Proc of American Control Conference, 1983, pp 1224-1229

Tong R.M. Analysis of fuzzy control algorithms using the relation matrix. Int. J. Man-Machine Stud. 1976; 8:679-686

Zadeh L.A. Calculus of fuzzy restrictions. In: Proc of US Open Seminar on Fuzzy Sets and Their Applications, Zadeh L.A., Fu K.S., Tanaka K. and Shimura M. eds., Academic Press, New York, 1975

CHAPTER 8
EXPERT SYSTEMS

Papermaking process operations involve multiple tasks under multiple criteria such as process quality control, operation optimization, fault detection, and emergency handling. The criteria applied in the operation are better product quality, higher production profit, improved equipment safety, and environment protection. Current process control and information systems mainly deal with normal situation and provide process operation information. How to efficiently utilize these information and make decisions for handling process operation problems is still heavily relied on process operators' experience.

The control system design for papermaking process also requires the systematic coordination of a multitude of tasks. Each of these tasks has many facets dealing with data, numerical algorithms, decision-making procedures, and human interaction to provide expertise knowledge. In most cases, it is a very complex and not a completely understood process particularly since it is abstract and requires creativity. It is largely a trial and error process, in which human design experts often employ heuristics or rules of thumb to solve their problems. The Computer Aided Design (CAD) techniques have evolved a new generation of design techniques. However, the present CAD techniques can only be used to deal with control algorithms, and to solve numerical analysis and simulation problems in the design process. As a new technological frontier, intelligent system technology has been extensively applied to control system design (James, 1987; Rao, et al, 1988; Sheu, 1987; Birdwell, 1987; Birky, et al., 1988; Rao, 1991).

Expert systems, as a subfield of artificial intelligence (AI), are being effectively applied to a variety of engineering problems. In this chapter, we will give a brief introduction to expert system in Section 8.1. A new technique using AI methodology for process control system design, namely Integrated Distributed Intelligent Design System (IDIDS), is proposed in section 8.2. The application

215

to paper machine headbox control system design is presented in Section 8.3. The conclusions are drawn as the last section.

8.1   Introduction to Expert Systems

An expert system is an interactive computer program helping an inexperienced user with expertise in a specific limited area of activity. Its capability of storing, manipulating, and searching symbolic and heuristic information allows us to code the human expertise into the computer program to solve the engineering problems at expert level (Donal, et al., 1991). The research and development in expert system has continued for many years, its efforts have produced three types of expert systems, they are: symbolic reasoning system, coupling system, and integrated distributed intelligent system (Rao, et al., 1990; Rao and Qiu, 1993).

The use of expert systems is increasing in industry in wide variety of applications. There are both tangible and intangible benefits that may be received from expert systems. They are listed as follows (Harris, et al., 1990):

Tangible benefits:

- Reduced costs;
- Increased production;
- Improved product quality;
- Reduced downtime;
- Faster decision making;
- Prevention of emergencies;
- More consistent reactions to process upsets;
- Retention of operator's expertise and knowledge.

Intangible benefits:

- Better use of experts' time;
- Improved personnel skill;
- Complementing currently installed technologies, improving technology acceptance, standardizing operating procedures, and providing consistent decision-making;
- Enabling the company to retain expertise;
- Evaluates existing expertise;
- Increased worker safety and improved working conditions.

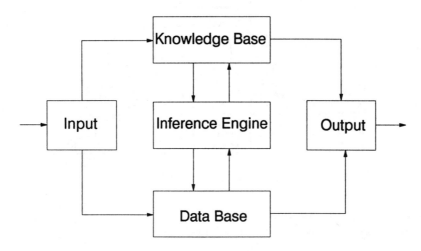

Figure 8.1: Structure of single symbolic reasoning system

8.1.1   Symbolic reasoning systems

A symbolic reasoning system is composed of database, knowledge base, inference engine and interface as shown in Figure 8.1.

Database has a special buffer-like data structure and holds the knowledge that is accessible to the entire system. Knowledge base contains two kinds of knowledge: public knowledge (general knowledge) and private knowledge (expertise knowledge). The expertise knowledge is described by a set of rules and frames. The typical production rule is described as "IF (condition), THEN (action)". Every production rule consists of a condition-action pair. The inference engine is an executor. It must determine which rules are relevant to a given data memory configuration, and select one to apply. Usually, this control strategy is called conflict resolution. The inference engine can be described as a finite-state machine with a three step cycle: matching rules, selecting rules and executing rules.

The segregation of the database, the knowledge base and the inference engine in an expert system allows us to organize the different models and domain expertise efficiently because each of these components can be designed and modified separately.

Presently, symbolic reasoning systems are extensively applied in the research of intelligent manufacturing systems. Among the successful AI applications, most expert systems are production systems. Production systems

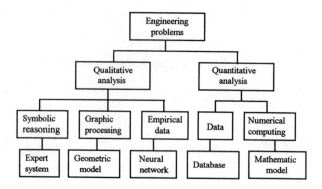

Figure 8.2: Qualitative and quantitative analyses

facilitate the representation of heuristic reasoning such that the systems can be built incrementally as the expertise knowledge increases.

8.1.2   Coupling systems

Many existing symbolic reasoning systems are developed for specific purposes. These systems can only process symbolic information and make heuristic inferences. The lack of numerical computation and coordination between applications limits their capability to solve the real engineering problems. In a process control environment, we require not only a qualitative description of system behavior, but also a quantitative analysis. The former can predict the trend of the change of an operating variable, while the latter may provide us with a means to identify the change range of the variable. We often use qualitative and quantitative analyses together in solving engineering problems (Figure 8.2). Usually, qualitative decisions are mainly based on symbolic and graphical information, while quantitative analysis is more conveniently performed using numerical information. Both methods often complement to each other.

Moreover, as a part of the accumulated knowledge of human expertise, many practical and successful numerical computation packages have been already available. Although artificial intelligence should emphasize symbolic processing and non-algorithmic inference, it should be noted that the utilization of numerical computation will make expert systems more powerful in dealing with engineering problems. Like many modern developments, artificial intelligence and its applications should be viewed as a welcome addition to the technology rather than a substitute for numerical computation. The coordination of symbolic reasoning and numerical computation is essential

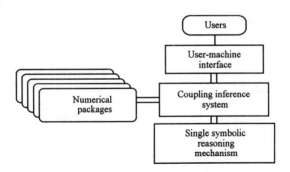

Figure 8.3: Structure of coupling expert systems

to develop expert systems. More and more, the importance of coordinating symbolic reasoning and numerical computing in knowledge-based systems is being recognized. It has been realized that if applied separately, neither symbolic reasoning nor numeric computing can successfully address all problems in control system design. The complicated problems cannot be solved by purely symbolic or numerical techniques (Kowalik and Kitzmiller, 1988; Rao, 1992).

Figure 8.3 distinguishes the coupling system from the symbolic reasoning system with the viewpoint of software architecture. So far, many coupling expert systems have been developed in various engineering fields to enhance the problem solving capacity of the existing symbolic reasoning systems. In IDSOC (Intelligent Decisionmaker for the problem solving Strategy of Optimal Control) (Rao, et al., 1988), a set of numerical algorithms to compute certainty factors is coupled in the process of symbolic reasoning. Another coupling system SFPACK (Pang, et al., 1990) incorporates expert system techniques in design package then supports more functions to designers.

Currently, not all of the expert system tools or environments provide the programming techniques for developing coupling systems. For example, it is very difficult to carry out numerical computation in OPS5 (Rao, et al., 1988). However, many software engineers are now building the new tools for coupling inference systems that will be beneficial to the artificial intelligence applications to engineering domains.

8.1.3 Integrated distributed intelligent system

The concepts of integrated distributed intelligent system (IDIS) were proposed by Rao, Tsai and Jiang (1987). Integrated distributed intelligent system

(IDIS) is a large-scale knowledge integration environment, which consists of several symbolic reasoning systems, numerical computation packages, neural networks, database management systems, computer graphics programs, an intelligent multimedia interface and a meta-system. The integrated software environment allows running programs written in different languages and communication among the programs as well as the exchange of data between programs and database. These isolated expert systems, numerical packages and modules are under the control of a supervising intelligent system, namely the meta-system. The meta-system manages the selection, coordination, operation and communication of these programs. The details about meta-system will be discussed in Section 8.2.4.

8.1.4   Development procedures of expert systems

The development of an expert system has five steps: identification, conceptualization, formalization, implementation and evaluation (Rao, 1992). The function of each step can be summarized as follows:

Identification: to define problem characteristics.

Conceptualization: to find concepts to represent domain knowledge.

Formalization: to design data structures to organize knowledge.

Implementation: to program knowledge structure into a computer program.

Evaluation: to test and modify the system.

Two successful strategies in developing expert systems have been summarized as: "Plan big, start small" and "Learning by doing" (Rao, 1992). Thus, the development for the real time intelligent control systems should be accomplished by three phases as follows:

Phase 1: Off-line system. The objective to develop an off-line expert system is to codify the knowledge, design and evaluate the system in off-line environment, for training or off-line supervisory control (Rao, et al., 1988]. A system at this phase is basically a simulation where the entries are performed by the user over a keyboard. Such a system is excellent for training purpose because it is a highly interactive system. Most academic intelligent systems are mainly developed at this phase. Their main function is to provide decision support to human operator based on process operation information.

Phase 2: On-line supervisory system. The objective to develop on-line supervisory system is to evaluate the intelligent system interfaces with human operator, process computer systems and hardware instrumentation. The intelligent system is physically connected to the actual process. The

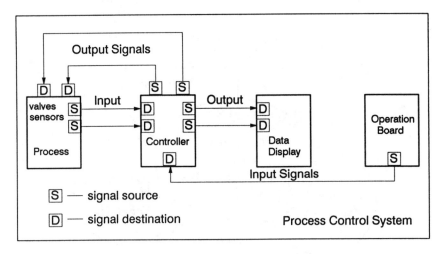

Figure 8.4: A typical process control system

process information is fully or partially processed by the intelligent system, and reports and suggestions are generated.

Phase 3: On-line closed loop system. An on-line closed loop expert system directly reads the inputs from the process and sends its outputs (actions) back to the process. However, human operators will be kept in the control loop. In most cases, the objective of these systems is to reduce the operators' routine tasks and improve operation safety and efficiency.

8.2 IDIS for Process Control System Design

8.2.1 Problem description

A typical process control system includes a process, a process controller, a control valve, measurement instrumentation, a process operator, as well as interconnection of them applied by sending signals (or messages), shown as Figure 8.4. Therefore, when designing a process control system, a designer considers not only controllers but also processes and operations. In other words, he should think the process control system as an integration of control, operation and process.

Process control system design can be divided into two aspects: conceptual design (creating and selecting models) and details design (analyzing and simulating these models). The former's objectives are to select appropriate

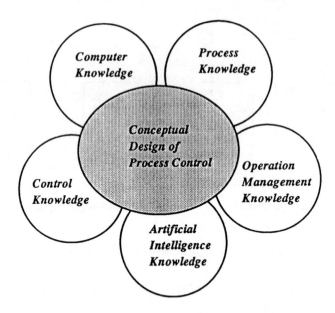

Figure 8.5: Knowledge model of conceptual design

controlled and manipulated variables, to choose a control configuration (e.g. feedback, feedforward etc.), to build control models (or control schemes), to study controllability, observability, reliability, operability and feasibility of control models, and to comprehensively decide the best model from all feasible models. The latter is to implement the process control system design in detailed based on the results of conceptual design. Clearly, the conceptual design is a creative activity and the most important decision making during the design. It needs to use various knowledge, techniques and methodologies from different disciplines such as computer science, control system, industrial process, operation management and artificial intelligence (see Figure 8.5).

A process control system has a hierarchical structure, which consists of many components and subsystems, such as process, controllers, measurement instruments, final elements and some associated units (such as recorder, monitor, alarm etc.), to facilitate numerous functions. A poor design may result in an incredible waste and bad system performance. The process control system design usually contains several design levels: system design, element design, and parameter design. In each design level, it may include three different designs: pattern design, creative design, and modifying design. The quality of the system performance depends on not only the quality of each component or subsystem but also the correct connection and coordination of components and/or subsystems. Thus, it often involves the cooperation of many experts from different disciplines.

Obviously, the conceptual design is a creative activity in a process control system design, in which the problems how to construct a system structure and how to choose elements (such as control element, measurement element, actuation element, etc.) according to the process dynamics, application requirements, as well as the knowledge of the element properties are addressed. As there exist more structures (patterns), a decision making process to generate a final solution is required. This is an ill-formulated problem, in which mathematical models are not amenable. Therefore, it is suitable for the application of artificial intelligence techniques (Rao, et. al., 1989; Lamont, et. al., 1987; Peng, 1989).

The elements and subpatterns in a pattern are called concepts, such as controller, measurement instrument, final element, and so on. The pattern design is to put several valid concepts together. Then, the attributes of each concept must be assigned. The process to determine attribute values is referred as parameter design. For a process control system, the conceptual design mainly consists of two parts: a pattern design (concepts combination) and a parameter design (attribute values determination). For instance, as pattern design, a controller should be selected as a PID controller, a PI controller, a PD controller, or a P controller, and therefore, in the phase of parameter design, the proportional gain, integral time and derivative time should be assigned the certain attribute values. The determination of attribute values not only implements concepts, but also provides data for analyses and evaluations.

8.2.2 Problem solving strategy

Problem solving strategy for design of process control system can be described as the following 5 stages (see Figure 8.6) (Rao, et al., 1992).

Stage 1 is a <u>Problem Definition</u> stage for design tasks (from application environment and objectives to functions). Functions to be used are chosen from the expertise function memory (it can be viewed as a part of knowledge base) according to the application environment and requirements provided by users (for example, optimal, decoupling or adaptive control etc.). The knowledge to define functions is shallow knowledge.

Stage 2 is an <u>Effective Conceptual Design</u> stage (from function to structure). The structures to execute functions are selected from structure memory (it can be viewed as another part of knowledge base). The communication

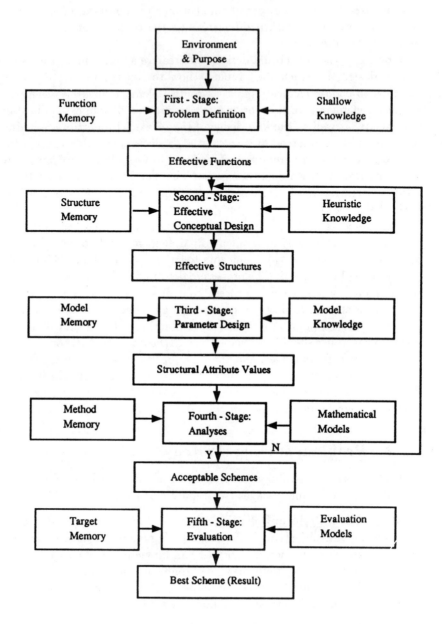

Figure 8.6: Problem solving strategy

between functions and structures is not an "one to one" mapping. Such a "multiple to multiple" mapping configuration indicates that a function could be realized with many different structures, and a structure may possess many functions. For example, the function for decoupling control can be implemented by different structures: state feedback, output feedback and feedforward. This "multiple to multiple" mapping makes the pattern design more complicated and diversified. If there are more than one design patterns, we need to select an optimal (or near optimal) one. In the case of no design patterns are available, new design techniques have to be used (since no existing structure can be used for the needed functions). If the existing design patterns fail to satisfy application requirements, the design has to be improved. The knowledge to formulate effective structure concepts is heuristic knowledge which can be represented by heuristic rules.

Stage 3 is a Parameter Design stage (from structure to parameter). At this stage, the detailed description of structures can be completed by using the design models stored in model memory according to the characteristics of effective structure concepts. The knowledge to determine structural attributes is deep knowledge, namely model knowledge, which is represented by object-oriented frames.

Stage 4 is an Analysis stage (from parameter to analysis). Because functions and structures share a "multiple to multiple" mapping configuration, numerous design schemes are usually produced. After parameters are given, all design schemes will be analyzed by selecting a numerical computation method (e.g. statistic analysis, optimization, and so forth) from method memory. Conventional CAD techniques can be utilized here.

Stage 5 is a final stage for comprehensive Evaluation (from analysis to evaluation). According to analysis data, a proper evaluation index system from index knowledge memory and a comprehensive mathematical model from evaluation models will be chosen to evaluate the selected practical scheme. Techniques of fuzzy mathematics and system engineering are used in evaluation.

Each stage in this problem solving strategy is very important. It combines numerical calculation (such as mathematical modeling, optimization and scheme analysis) with symbolic reasoning (knowledge representation and model handling as well as scheme evaluation) to accomplish the objectives in every stage.

8.2.3   System configuration

A good problem solving strategy must match a good program structure to ensure the quality and efficiency of the software system. The Integrated Distributed Intelligent System (IDIS) for process control conceptual design that we develop is an integrated intelligent system (see Figure 8.7), which consists of several symbolic reasoning systems and numerical computation packages. The module technique and meta-system architecture are used to implement IDIS. Each module performs different function. For example, task definition module provides a window to input information. User can define design tasks, application environment, purposes and specifications through the window. Also it contains: function design module, structure design module, several element design and parameter design modules, as well as analysis and evaluation modules. As a subsystem, each module may be written in different languages and be used independently. They are under the control and management of a supervising expert system, namely, a meta-system. This structure simulates the human being reasoning behavior in design process such that it can be used as a general framework for developing expert systems to accomplish conceptual design (Rao, et al., 1992). The following lines briefly describe the module's functions and characteristics.

(1) **Menu Management System:** It can guide users to select modules and to observe the performance of the modules. Since the menu system employs a tree structure (each subsystem has its own sub-menu) and an object-oriented programming technique for man-machine interface, users can select and run modules according to the contents on screen.

(2) **Task Definition Module:** It is a window to input information. Users can define design tasks, application working environment, purposes and specifications through the window. The information that is normally available at the initial stage of a design problem is given such as performance and size of process equipment.

(3) **Function Concept Design Module:** It functions as the first step in the problem solving strategy. Its purpose is to further expand facts and information in the knowledge bases due to the existing specifications to reduce interactive contents, and to determine some the global variables (parameters) that might be shared by many subsystems. In general, global variables are not equally well understood, which need to comprehensively consider various situation to happen. It is obvious that the entire process control must be considered when determining these global, however, it is also very difficult for a new designer to do. In addition, a user never ensures that he/she receives

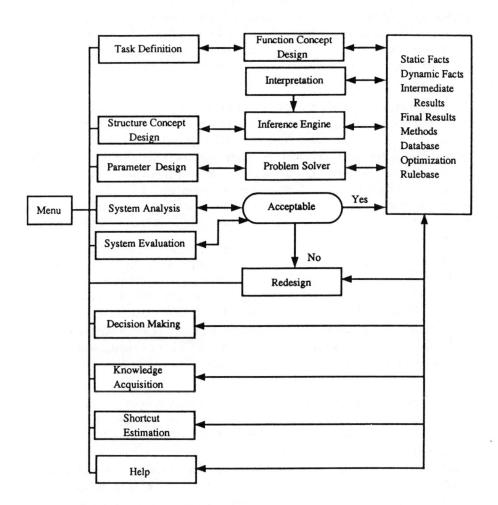

Figure 8.7: Control structure of the software system

the right and complete information. If some important data are missed, the module should supply these data using domain specific knowledge.

(4) **Structure Concept Design Module:** It can select the types of scheme and potential structural configuration components to satisfy the functions and key constraints. This module works in the second step and provides parameter design with a variety of features. Its knowledge base includes experts' experience and heuristic knowledge which are represented as heuristic rules. Inference employs the constraint reasoning.

(5) **Parameter Design Module:** It performs its function that is similar to the third step of design process. With the structures obtained from concept design and the facts provided by users, attribute values (parameters) of the systems are determined. Parameter design usually deals with local variables by shortcut calculations. Local variables are well understood by engineers, because these variables only associate with a subsystem, rather than the overall system or other subsystems. The knowledge of parameter design is expressed by the data structure of object-oriented frame that is operated by the problem solver.

(6) **System Analysis Module:** It functions in the fourth step. Based on the results from pattern design, element design and parameter design, this module analyzes each preliminary design scheme and provides data for evaluation. There are two paths in the module. If there exist more satisfactory specifications, then the system selects next module. Otherwise, the local redesign is required. In addition, the module can call the existing analysis and simulation software packages such as the time domain, frequency domain as well as state space calculations, etc.

(7) **System Evaluation Module:** This module is equal to the fifth step of problem solving (comprehensive evaluation). Its purpose is to evaluate comprehensive functions. In other words, all schemes entering the evaluation module are practical ones that are different from each other only in quality. The evaluation uses the comprehensive evaluation models, fuzzy mathematics and system engineering techniques. Indices (or targets, such as operability, controllability, observability, etc.) and weights are selected by domain experts. There are two paths in the evaluation module. If the evaluating results satisfy the specifications provided by users, the information is sent to the Decision Making Module. Otherwise, a global redesign will be performed.

(8) **Decision Making Module:** In general, it is very often to generate multiple schemes as design result. During a process control system design, the index system solicits various opinions from different domain experts. Thus,

the conflicting solutions are generated from the different experts opinions. The decision making module will pick up a best scheme among these schemes.

(9) **Interpretation Module:** It connects with the inference engine and provides the interpretation for concepts and reasoning paths in order to help users understand items and concepts, then to manager the system.

(10) **Inference Engine:** Inference engine usually performs specific tasks and formulates knowledge representation. Obviously, in order to solve the large hybrid problem in control system conceptual design, a simple inference engine that provides only one reasoning technique is unsatisfactory. The integration of different inference mechanisms are very demanding in solving the real world problems.

(11) **Static Facts Base:** SFB stores task definitions and specifications provided by users. The information in SFB contains the essential conditions (constraints) for functions and structure design. The facts in SFB are expressed with vector lists.

(12) **Dynamic Facts Base:** In a reasoning process, users must continuously provide the more detailed facts and data which are stored in DFB. In our integrated software system, DFB only associates with inference engine and supports the interpretation Module.

(13) **Method Base:** It records the structure descriptions for all problem solving methods. Each parameter or structure attribute has its own specific methods to be generated through reasoning, table look-up, analogy, calculation and so on. MB is operated in many different ways. When needed to handle a new parameter, users may add description to MB. All methods are described with an object-oriented frame.

(14) **Optimization Base:** It provides several optimization techniques, such as linear optimization, nonlinear constraint optimization, discrete optimization.

(15) **Help:** The help module brings up a context sensitive help window explaining each of the system commands. The help window can be selected at any time from any module.

The configuration and software platform for developing expert systems of process control system conceptual design discussed above has the following features:

1. It is an integrated distributed intelligent system (IDIS) with a userfriendly interactive environment, which can be used to develop concept

design (scheme design) expert system for control systems or their components.

2. It provides the problem solving strategy for design types with multiple objectives or uncertain goals.

3. Its system configuration is an open structure. With different requirements, users can modify the system at any level of system. Compared with closed-structure expert system environments, this open-structure system is more flexible and easier to integrate various expert systems and existing numerical software packages.

4. A variety of knowledge representation techniques are provided. Each knowledge representation technique is associated with an inference engine. Users can choose these techniques whenever needed.

5. It combines numerical calculation and symbolic reasoning as well as graphic simulation.

6. Its decision support subsystem consists of several modules for analysis, evaluation, and decision making. It can evaluate all schemes and select the best one.

7. It can accomplish the redesign of optimization backtrack.

8. The system provides a good interface for the analysis module.

9. Since the knowledge of reasoning, methods of analyses, and formula of calculation are separated from control structure and inference engines, it is easy to modify and manage knowledge base.

8.2.4   Software implementation platform

Based on the configuration discussed above, the software platform for developing IDIS has been implemented by using meta-system architecture. Meta-system is a supervisory intelligent system, which has its database, knowledge base, and inference engine. The main tenets of our view about the meta-system are (Rao, et. al, 1989; 1990):

(1) Meta-system is the coordinator to manage all symbolic reasoning systems, neural networks, numeric computation packages and computer graphics programs in an integrated distributed intelligent system. The hierarchy of the integrated distributed intelligent system as described in Figure 8.8 indicates

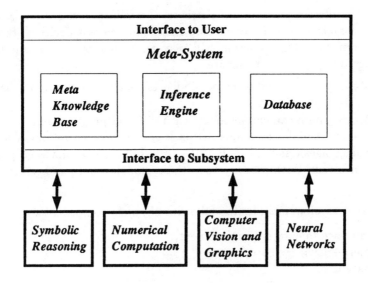

Figure 8.8: Hierarchy of IDIS

that the different type of information (symbolic, graphic and numerical) are utilized and processed together. For numerical information, we can use algorithm to handle equation-oriented data, and use neural network to deal with empirical data that can not use the algorithm to handle. Even in the branch of symbolic information, many different languages and tools may be applied to build individual symbolic reasoning systems, and so does the numerical computation branch.

(2) Mate-system distributes knowledge into separate expert systems and numeric packages so that the IDIS can be managed effectively. Therefore, the knowledge bases of all expert systems are easier to be modified by end users.

(3) Meta-system is the integrator which can easily acquire new knowledge. It provides us a free hand in integrating and utilizing new knowledge. Each of the modules in IDIS is separated with one another and only ordered strictly with the meta-system. When a new expert system or numerical package is to be integrated into the IDIS, we just have to modify the interface and the rule base of the meta-system, while other modules or programs are still kept unchanged.

(4) Meta-system can provide a near optimal solution for conflict solutions and facts among the different expert system. Knowledge sources are distributed into various distinctive modules that correspond either to different

tasks or to different procedures. However, the conflict usually appears since different domain expert systems could make the different decisions based on different criteria. This function will play an even more important role in knowledge integration in the future.

(5) Meta-system provides the possibility of parallel processing in the IDIS.

(6) Meta-system can communicate with different programming languages. In IDIS, some modules or numerical packages are written by ourselves with different languages (such as C, BASIC, etc.), but a few commercial packages are also involved. With meta-system, these modules and packages can work together.

The meta-system includes the following six main components:

(1) interface to external environment,
(2) meta-knowledge base,
(3) database,
(4) inference mechanism,
(5) interface to internal subsystems,
(6) static blackboard.

The meta-system is implemented in C language environment. There are four main reasons to use C as implementation language. First of all, C language is versatile in both numeric computation and symbolic manipulation. Its capability to handle numerical operation is much more powerful than Lisp, Prolog, OPS5 and other expert system development tools. It is also superior to FORTRAN and BASIC in terms of symbolic operations. This advantage makes C easier to integrate different forms of knowledge. Secondly, C language possesses merits of both high level and low level languages such that it is very flexible and convenient for program coding and control on hardware, especially under UNIX operating system that is developed in C. Thirdly, C language can easily access other language environment by interfaces written in mixture of C and the assembly language. Finally, C++ is object-oriented programming language and is an extension of C language. Details about the implementation of meta-system can be found in references (Rao, et al., 1989; 1993).

8.3 Application to Headbox Control System Design

The application of IDIS to a paper machine headbox control system design is presented as a simple example.

The paper machine headbox control system is commonly used in many pulp and paper mills as a complementary package to control basis weight and moisture. In general, the benefits from designing a good headbox control system include: improved wet end stability, better and more uniform formation, automatic wet end start-up and speed change capability, reduction of variations in machine direction and cross direction tensile and tear ratio, and other final sheet quality characteristics. The headbox control schemes are often obviously varied according to the headbox type, the stock delivery system as well as control system requirements.

As shown in Figure 8.9, the IDIS for headbox control system design mainly consists of Meta-COOP which functions as meta-system, control scheme design expert system, controller type design expert system and controller action direction expert system, as well as a numerical analysis and simulation package, namely PCET. Among them, Meta-Coop is developed in C^{++} (more information can be found in Chapter 10), the control scheme design expert system is implemented in PCPLUS (commercial expert system development tool), the controller type design and action direction selection expert systems are developed in C, and PCET is a commercial package.

The main menu contains [**Introduction**] [**Knowledge base**] [**Inference engine**] [**File management**] [**Datebase**] [**System design**] and [**Help**]. As an example, under [**Knowledge base**], several functions such as ⟨Add⟩ ⟨Delete⟩ ⟨Modify⟩ and ⟨View⟩, are provided. User can add, delete, modify or view knowledge base by choosing the corresponding function. The production rules and structured objects are used as knowledge representation techniques. For easy maintenance and high speed of rule chaining, the rules are grouped into different rule modules. At a particular stage of the design process, only the appropriate rule base will be loaded in. Thus, it can avoid the exhaustive search when the knowledge base becomes large.

As an example of knowledge representation, a rule used at the evaluation stage is illustrated as follows:

> *IF   closed − loop system is stable;*
> *AND   steady − state response is satisfactory;*
> *AND   dynamic performance is good.*
> *THEN the designed control system is acceptable.*

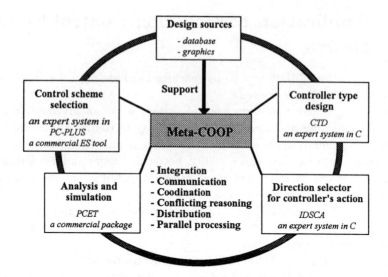

Figure 8.9: Architecture of IDIS for control system design

Similarly, under the menu of [**System design**], ⟨Problem Definition⟩ ⟨Scheme Design⟩ ⟨Controller Design⟩ ⟨Parameter Design⟩ ⟨Direction Selector⟩ ⟨System Analysis⟩ and ⟨System Evaluation⟩ are provided as the sub-menu. Their functions are briefly described below.

⟨Problem definition⟩

It will ask user to provide some basic and specific information, such as the type and size of the headbox, the variables need to be controlled, the steady state operation conditions, the stock delivery system, and the control quality requirements, as well as the constraints. All information are stored in *Static Facts Base*.

⟨Scheme Design⟩

Based on the information provided by user and the system function, the control system scheme is selected from *Structure Memory*. The main control problems encountered in the headbox control are to maintain total pressure and level in the headbox, and to maintain the rush/drag ratio when wire speed change. Usually the total head is controlled by adjusting thin stock flow rate while level is controlled by adjusting air pad pressure. In some cases, the fluctuations in stock flow may be caused in many ways, such as undamped pulsations from pumps, rotating screens, etc. Therefore, the cascade control may be used to eliminate or minimize such disturbances before they reach the

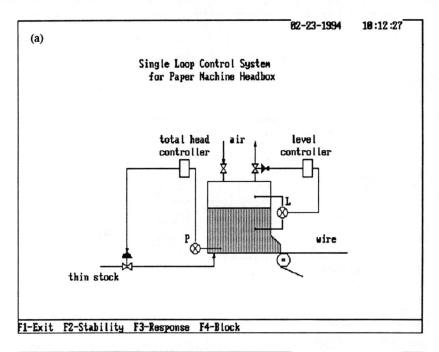

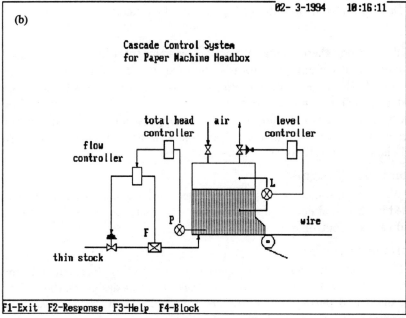

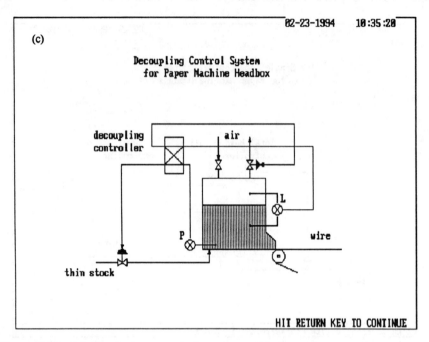

Figure 8.10: (a) Single-loop control system. (b) Cascade control system and (c) Decoupling control system for paper machine headbox

headbox. Since there exists interaction between total head and level control systems, the decoupling control may be used in order to improve the control quality. If the high control performance is required, the advanced control strategies such as nonlinear control, robust control, optimal control and others may be used. For a case study, only single loop, cascade and decoupling control schemes are built in the *Structure Memory*, as shown in Figure 8.10 a-c. Figure 8.11 shows the results of scheme design.

⟨Controller Design⟩

What type of controller should be used in designed control scheme is selected in this module. Figure 8.12 shows the controller type design for total head cascade control system.

⟨Parameter Design⟩

By choosing an appropriate algorithm from *Method Base*, the parameters of the controller are designed at this stage. Several design methods for PID controllers and decoupling controllers are employed in *Method Base*. Other numerical computation packages for process dynamic analysis, system stability

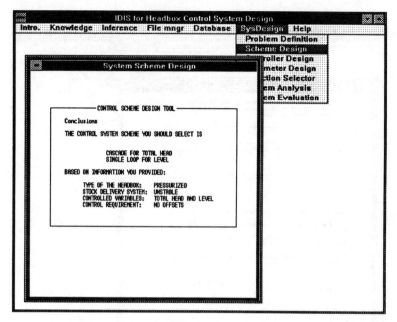

Figure 8.11: A slide in the scheme design

analysis and time domain response analysis etc., are also available from *Method Base*.

In this example, the cascade controllers for total head control system is designed. The block diagram of the system is shown in Figure 8.13. The process model can be described by following transfer functions:

$$G_{p2}(s) = \frac{1}{0.18s + 1}$$

$$G_{p1}(s) = \frac{1.75}{2.6s + 1}$$

Through parameter design, we obtain the results:
for secondary loop

$$K_{c2} = 2$$

for primary loop

$$K_{c1} = 1.5, \qquad T_i = 1$$

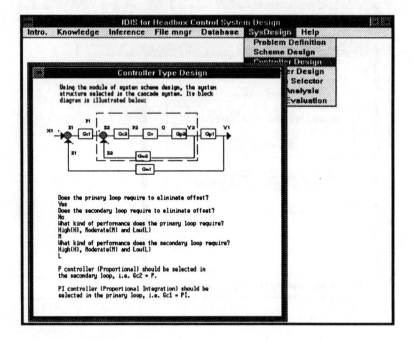

Figure 8.12: A slide in the controller design

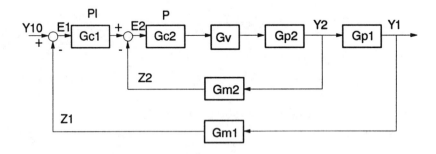

Figure 8.13: Block diagram of the cascade control system

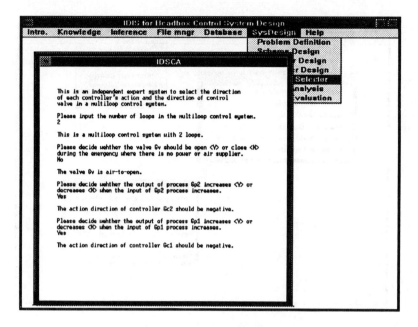

Figure 8.14: A slide in direction selector

⟨Direction Selector⟩

Choosing correct direction of controller's action is very important for industrial environment. This module can help user to select valve type (air-to-open or air-to-close) and controller' action direction in the multi-loop control systems. Figure 8.14 shows the result of valve type and controllers' direction selected for total head cascade control system.

⟨System Analysis⟩

After all parameters are given, the designed system will be analyzed at this stage. Figure 8.15 shows the different options for system analysis. *Process Dynamic Analysis* gives user the process characteristics of the open loop system. *System Stability Analysis* provides the information about the designed system's stability. The control system can be designed by *Root Locus Design* and *Frequency Domain Design. Time Domain Response Analysis* shows the performance of the designed system. *Controller Tuning* provides the tools for adjusting controller parameters. If the system satisfies the performance requirements provided by user, then the design results are sent to ⟨**System Evaluation Module**⟩. Otherwise, other control strategy has to be selected.

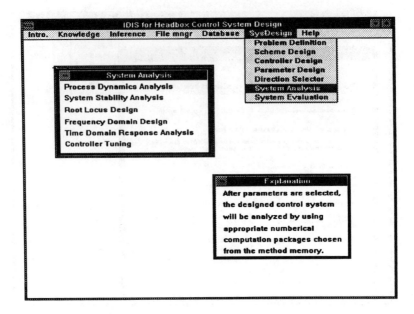

Figure 8.15: Options for system analysis

⟨**System Evaluation**⟩

Comparing the system analysis results with control quality requirements, the final decision is made. In this example, the effective control system for a pressurized headbox is cascade control for total head and single loop control for headbox level. The simulation results of designed cascade control system performance is presented in Figure 8.16.

This intelligent system for process control system design is a large knowledge integration environment. It provides the following features:

1. It is an integrated intelligent design environment. Its systematic functions include conceptual design, dynamic simulation, performance analysis, and system evaluation.

2. Its modularity enables system configuration so flexible that the knowledge base is easy to be expanded and modified by the end users, rather than the original developers.

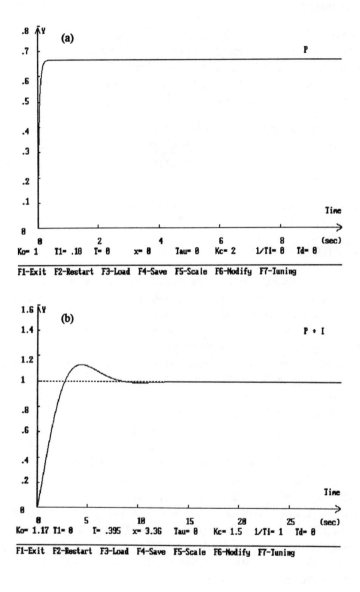

Figure 8.16: Simulation results of the cascade system. (a) secondary loop; (b) primary loop

3. It can accomplish both multi-objective design tasks and uncertain multi-criteria design tasks.

4. Integrated distributed intelligent system concept provides an open structure to organize this intelligent system. Such a configuration allows the user to modify knowledge base at any level of the system. In this intelligent system, numerical computing process is coupled with symbolic reasoning to enhance system capability.

5. Several different problem solving strategies, reasoning mechanisms and knowledge representation techniques are implemented in this system. A user can select the suitable knowledge processing techniques to deal with his (or her) specific design problem.

6. It interfaces with the commercial software packages.

8.4   Conclusions

In this chapter, expert systems, as the most common used technology of artificial intelligence (AI) in engineering applications, were introduced. Based on the software architectures and system functions, we classified the current expert systems into three types, i.e., single symbolic reasoning system, coupling system and integrated distributed intelligent system.

The IDIS for process control system design was presented. The methodology for developing IDIS can be used as a general methodology to develop integrated distributed intelligent systems for process design, process control, planning, scheduling, and other activities.

In IDIS, since the knowledge of reasoning, methods of analyses, and formula of calculation are separated from control structure, e.g. problem solver and inference engine, it is easy to modify and manage knowledge bases. The system configuration is an open structure. With different requirements, user can modify the system at any level of the system.

The configuration of IDIS discussed here can be used for process control system design but also for other design activities. The meta-system can coordinate symbolic reasoning and numeric computing processes, distribute knowledge sources, find a near optimal solver for conflicting solutions, as well as provide parallel processing capability.

8.5 References

Birdwell J.D. An expert system can aid in the evolution of a design methodology. Proc of American Control Conference, Minneapolis, MN., 1987, pp 541-546

Birky G.R., McAvoy T.J. and Modarres M. An expert system for distillation control design. Computers. Chem. Engng 1988; 12:1045-1063

Donal R., Windon Jr., Deborah F.C. and Joseph G.M. Integrating the art and science of paper making in an expert system. TAPPI Journal 1991; 9:85-89

Harris C.A., Sprentz P., Hall M. and Meech J.A. How expert systems can improve productivity in the mill. Pulp and Paper Canada 1990; 91(11): 29-34

James J.R. A survey of knowledge-based systems for computer-aided control system design. Proc of the American Control Conference, Minneapolis, MN., 1987, pp 2156-2161

Kowalik J.S. and Kitzmiller C.T. Coupling symbolic and numeric computing in expert systems, II Ed. Elsevier Science Publishers, New York, 1988

Lamont G.B. and Schiller M.W. The role of artificial intelligence in computer aided design of control system. Proc of Conference on Decision and Control, Los Angeles, CA., 1987, pp 1960-1965

Pang G. Knowledge engineering in the computer-aided design of control system. Expert Systems 1989; 6(4):250-262

Rao M. Integrated systems for intelligent control. Springer-Verlag, Belin, 1991

Rao M. Challenges and frontiers of intelligent process control. Engineering Application of Artificial Intelligence 1992; 5(6):474-481

Rao M., Jiang T.S. and Tsai J.P. IDSCA: An Intelligent Direction Selector for the Controller's Action in multiloop control systems. International Journal of Intelligent Systems 1988; 3:361-379

Rao M, Jiang T.S. and Tsai J.P. Combining symbolic and numerical processing for real-time intelligent control. Engineering Application of Artificial Intelligence 1989; 2:19-27

Rao M, Jiang T.S. and Tsai J.P. Integration strategy for distributed intelligent system. J. of Intelligent and Robotic Systems 1990; 3:131-146

Rao M. and Qiu H. Process control engineering. Gordon & Breach Science Publisher, Langhorne, Pennsylvania, USA, 1993

Rao M., Wang Q. and Cha J. Itegrated distributed intelligent systems in manufacturing. Chapman & Hall, London, 1993

Rao M., Tsai J.P. and Jiang T.S. An intelligent decision-maker for optimal control. Applied Artificial Intelligence 1988; 2:285-305

Rao M., Ying Y. and Wang Q. Integrated distributed intelligent system for process control system design. Engineering Application of Artificial Intelligence 1992; 5(6):505-518

Sheu P. and Kashyap R.L. Designing control systems with knowledge. Proc of American Control Conference, Minneapolis, MN., 1987, pp 941-946

CHAPTER 9
MODELING VIA ARTIFICIAL NEURAL NETWORK

In this chapter, artificial neural network technology is applied to predict the basis weight and moisture content to improve the paper product quality. Historical data from a paper production company in Canada are analyzed and applied to train a multilayer feedforward backpropagation network. Considering that generalized descent method, which is a typical optimization algorithm in backpropagation, has some major drawbacks, a conjugated gradient method is proposed for training neural networks. The results have shown that the neural network gives accurate paper quality prediction. The application of artificial neural network helps us to gain a better understanding of dependence of quality variables on the operating conditions and to overcome large time-delay in paper machine control systems.

9.1   Introduction

Paper making is a very complex process. Many parameters affect product quality and production profits. The knowledge concerning the operation includes complex technologies from different areas. In addition, on-line measurement of many important variables is either unreliable or impossible due to sensor technology limitations. In many cases, control is dependent on unreliable, noisy or manually gathered data. Under these conditions, even experienced operators find it difficult to deal with operations such as quality control and operation optimization.

To alleviate these problems, many paper production companies have installed the state-of-the-art control and information systems. However, some important process variables, such as basis weight and moisture content, may not be accurately measured on-line. These crucial quality variables can only be measured manually, which is not conducive to efficient process control.

In other cases, even though on-line measurements exist, the processes have such long dead-times that the measurement is inadequate for proper control. For example, the time delays from the basis weight valve in headbox section to reel basis weight and from steam pressure to moisture content are as large as twenty minutes. After the changes in the quality variables present, it is already too late to take corrective actions. With adequate modeling, for the first case the various process outputs could be estimated, and the information could be presented to operators/control systems resulting in better decisions/control; for the second case, the prediction of various process outputs would be well in advance of the actual result being produced, so corrective actions could be taken proactively.

Attempts to solve these quality prediction problems using traditional numerical methods and statistical techniques have not been completely satisfactory. One reason is that the process mechanism is very complex and not well known. It is impossible to get all rigid numerical models. Another reason is that the process models obtained from traditional methods such as system identification are usually linear approximations about a chosen steady state operating condition. However, as pointed in the previous chapters, the characteristics of paper machines are highly nonlinear and very complex. When the actual operating condition moves away from the one on which we linearize the process model, the discrepancy between the obtained linear model and the actual process characteristics will become much more significant. The prediction based on the linear model will give inaccurate results.

On the other hand, most of the nowadays pulp and paper companies have applied distributed control systems and information management systems. They have vast amount of historical process data charting the system. Neural network is trained from data and examples. These stored historical data can be used to develop a neural network model that describes the input/output behavior of the process. Therefore, artificial neural network technology can be used to solve the above problems for which the traditional numerical model-based systems are not amenable.

In order to improve product quality, decrease raw material and energy consumption, increase production, and reduce environmental pollution, there is a need to develop neural networks for accurate prediction of product quality.

After two decades of near eclipse, interest in artificial neural network has grown rapidly over the past decade. The renewed attention on neural network dues both to theoretical advances which make feasible systems with generalized decision making capabilities, and hardware advances which make the practical implementation of large neural network structures possible.

A neural network can learn and adapt itself to inputs from real processes, thus allowing representation of complex engineering systems that are difficult to model with traditional physical relations. In addition to self- organizing capability, another advantage of neural network is the parallelism inherent in the neurocomputing architecture. This allows neural network system to be implemented using highly parallel hardware to achieve real-time performance. Several factors motivate the use of neural networks:

- They are capable of extracting essential characteristics from vast amount of inputs containing irrelevant information.
- They are able to learn from experience, whereas most other techniques rely on pre-specified algorithms.
- They have ability to generalize from previous examples to new ones.
- They exhibit a greater degree of robustness and fault tolerance.
- They can respond to sensor inputs quickly.

Many areas of industrial process operation, such as process modeling, process control, fault diagnosis, operation optimization can take advantage from the above properties.

In the past a few years, considerable researches have been done in the application of artificial neural network in process industry. Behnam (1988) introduced neural network in the control. Hoskins and Himmelblau (1988) applied neural network model to represent the knowledge in chemical engineering. Process modeling is one of the areas in which neural network is wide used. Bhat and McAvoy (1990), Haesloop and Holt (1990) and Kim and his coworkers (1992) used neural network for dynamic modeling and control of chemical processes. neural network is also widely applied in process fault diagnosis (Venkat and King, 1989).

In this chapter, process modeling and quality prediction using neural network are presented. It is shown that neural network offers the opportunity to directly model nonlinear processes. With a more accurate nonlinear model, the plant-model divergence mentioned above may be reduced. Neural network modeling is an area of such concern with a non-programmed adaptive approach to processing engineering data systems. Neural networks can adaptively develop transformations in response to their environment. In contrast to programmed computing, neural networks are able to develop a mapping function from examples of that functions operation.

9.2 Fundamentals of Artificial Neural Network

A neural network is nothing more than a computational system that performs brain like functions simply because it was modeled similar to the human brain. It is composed of highly interconnected simple processing elements, called artificial neurons, in parallel. They are organized in patterns similar to biological neural networks. Due to its structural and functional resemblance to a biological neural network, it exhibits a number of characteristics of the human brain, for example, learning from experience by modifying its behavior in response to its environment, generalizing from previous examples to new ones as a result of its structure and abstracting essential characteristics of a set of inputs containing irrelevant data.

The idea of building an artificial neural network was originally conceived as an attempt to model the biophysiology of the human brain, in other words, to understand and explain how the human brain operates and functions. Along with the progress in neuroanatomy and neurophysiology, psychologists were developing models of human learning in order to produce computational systems that perform brain like function.

In 1949, Hebb proposed a learning law that proved to be the most successful model and became the starting point for neural network training algorithms. It showed scientists how a network of neurons could exhibit learning behavior.

During 1950s and 1960s, a group of researchers combined these biological and psychological insights to produce the first neural network, which was initially implemented as an electronic circuit. Later it was converted to the more flexible medium of computer simulation. Then many researchers including Marvin Minsky developed networks consisting of a single layer of artificial neurons call perceptrons and applied them to such diverse problems as weather prediction, electrocardiogram analysis, artificial vision, etc. At that time, they thought that reproducing the human brain was only a matter of constructing a large enough network (Wasserman, 1989). However, networks failed to solve problems superficially similar to those they had been successful in solving. As a respected senior scientist in this field, Minsky carefully applied mathematical techniques and developed rigorous theorems regarding network operation.

In 1969, Minsky published a book titled PERCEPTRONS (Minsky and Papert, 1969). In this book, he proved that the single-layer networks can not theoretically solve many simple problems, including the function performed by a simple exclusive-or gate. Even he was not optimistic about the potential for progress. So discouraged researchers left the field for areas of great promise.

Government agencies redirected their funding and neural network lapsed into obscurity for nearly two decades. Nevertheless, a few dedicated scientists continued their efforts. Gradually, a theoretical foundation emerged, upon which the more powerful multilayer networks of today are being constructed.

In 1986, Rumelhart and other researchers invented backpropagation which provides a systematic means for training multilayer network, thereby overcoming limitations presented by Minsky (Rumelhart, et al., 1986). Although a broad range of neural network architectures and learning paradigms are available, the backpropagation algorithm for multilayer feedforward network is the most popular approach for current engineering problems because of its simplicity and powerfulness.

A human brain is known to contain over a hundred billion neurons which are considered to be computing elements. These neurons communicate throughout the body by way of nerve fibers that make perhaps one hundred trillion connections. A neuron is the fundamental building block of the nervous system. It is called a cell similar to all cells in the body. However, certain critical specializations allow it to perform all of the computational and communicational functions within the brain. The neuron consists of three sections: the cell body, the dendrites, and the axon, each with separate but complementary functions. Dendrites extend from the cell body to other neurons where they receive signals at a connection point called a synapse. On the receiving side of the synapse, these inputs are conducted to the cell body. There they are summed, some inputs tending to excite the cell, others tending to inhibit its firing. When the cumulative excitation in the cell body exceeds a thresh hold, the cell fires, sending a signal down the axon to other neurons. Even though this basic functional outline has many complexities and exceptions, most neural networks model only these simple characteristics.

The artificial neuron was designed to mimic the first-order characteristics of the biological neuron. Figure 9.1 represents an artificial neuron model. In essence, a set of inputs labeled X_1, $X_2, \cdots$, X_n is applied to the artificial neuron. Each input represents the output of another neuron. Each signal is multiplied by a corresponding weight W_1, W_2, $\cdots$, W_n before it is applied to the summation block, labeled $\sum$. Each weight corresponds to the strength of a single biological synaptic connection. The summation block, corresponding roughly to the biological cell body, adds all of the weighted inputs algebraically to determine the activation level of the neuron. The summed signal S is usually further processed by an activation function F to produce the neuron's output signal Y (Figure 9.2). If the activation function F compresses the range of S, so that Y never exceeds some low limits regardless of the value of S, F is called a squashing function. The squashing function is often chosen to be sigmoidal

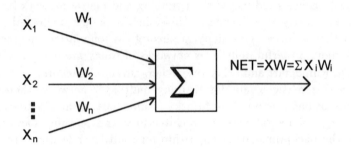

Figure 9.1: Artificial neuron model

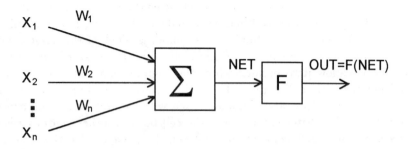

Figure 9.2: Artificial neuron with activation function

function that is continuous and strictly monotonic as shown in Figure 9.3. This function can be expressed mathematically as $F(S) = 1/(1 + e^{-S})$. Another commonly used activation function is the hyperbolic tangent, $F(S) = tanh(S)$.

A typical representation of the three-layer feedforward network is shown in Figure 9.4. It is composed of many interconnected processing units or neurons (N_0 neurons in the input layer, N_1 neurons in the hidden layer and N_2 neurons in the output layer) organized in successive layers. The analysis and design of any multilayer neural networks can be done similarly. The neuron j in the layer k first computes the weighted sum of the N_{k-1} inputs

$$S_{j,k} = \sum_{i=1}^{N_k-1} W_{i,j,k} X_i \tag{9.1}$$

where X's and W's represent the given input and its associated weight, respectively. Then it outputs a nonlinear function of the sum in (9.1), which

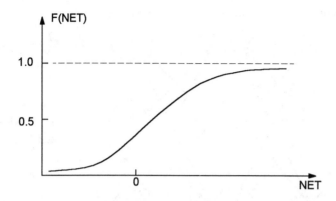

Figure 9.3: Sigmoidal function

serves as an input to the next layer

$$Y_{j,k} = F(S_{j,k}) \tag{9.2}$$

Every neurons except those in the input layer functions in the same way. In the first or input layer, neurons do not perform any computation but simply distribute their inputs to all neurons in the next layer.

The only unknowns in this representation are these associated weights between layers. In order to use the neural network model, these weights should be determined first. The evaluation of appropriate weights is called learning or training. The objective of learning is to train the network so that application of a set of inputs produces the desired or at least consistent set of outputs.

9.3 Backpropagation Learning Paradigm

Even though a broad range of neural network architectures and learning paradigms are currently available, the backpropagation algorithm for multi-layer feedforward neural networks is the most common approach for engineering applications mainly because it is simple and powerful.

In the backpropagation learning paradigm, the determination of weights is accomplished by sequentially applying a set of inputs and desired outputs to the network, while adjusting network weights according to a predetermined procedure. During training, the network weights gradually converge to values such that each input data set produces the desired output.

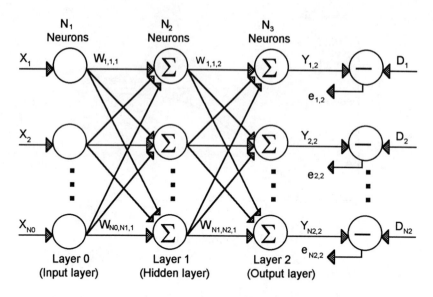

Figure 9.4: Three-layer backpropagation network

The objective of the backpropagation learning paradigm is to minimize the overall error, E, between the desired and actual outputs

$$E = \sum_{l=1}^{p} \sum_{k=1}^{N_2} e_{k,2}^{(l)} \tag{9.3}$$

$$e_{k,2}^{(l)} = \frac{1}{2}(D_k^{(l)} - Y_{k,2}^{(l)})^2 \tag{9.4}$$

where D's and Y's represent the desired outputs and the actual outputs, respectively. p represents the number of data set used to train the network. The sum of errors between the desired outputs and actual outputs is required to be smaller than the given error tolerance for the network to identify the system using the input-output data.

The backpropagation is known as a supervised synchronous learning paradigm. It is supervised because for every input presented for learning, the expected or correct corresponding output is available so the network can modify its weights accordingly by applying the gradient descent method. It is synchronous because all the weights are modified at each learning step. The network is expected to learn a number of data patterns composed of inputs and associated outputs.

The basic strategy behind the standard backpropagation algorithm is to perform gradient descent in parameter space based on the individual errors between a set of network mappings and the corresponding set of desired mapping examples

$$W_{i,k,j} = W_{i,k,j} + \eta \Delta W_{i,k,j} \qquad (9.5)$$

Before starting the training process, all of the weights are initialized at small random values. This ensures that the network is not saturated by large values of the weights, and prevents certain other training pathologies. The first pattern is then presented to the network. Learning takes place in two successive steps: forward and backward passes.

During the forward pass, as described in the previous section, each neuron i in layer j computes the weighted sum of its inputs $S_{i,j}$ and outputs $Y_{i,j}$ as in (9.1) and (9.2), respectively. This process is called forward because it takes place in layer j-1 before layer j and therefore propagates from the input layer to the output layer. At this point the activations or outputs of all the neurons of the network are available.

During the backward pass, the output error is backpropagated in the reverse direction to minimize the error, $E_{i,2}$, between the desired and actual outputs

$$E_{i,2} = (Y_{i,2} - D_i)F'(S_{i,2}) \qquad for \ 1 \leq i \leq N_2 \ (output \ layer) \qquad (9.6)$$

where $F(\cdot)$ denotes the derivative of the activation function. For each of the N_1 hidden units the error is computed as

$$E_{i,1} = F'(S_{i,1}) \sum_{k=1}^{N_2} E_{k,2} W_{i-k,1} \qquad for \ 1 \leq i \leq N_1 \ (hidden \ layer) \qquad (9.7)$$

Once the errors has been computed, the weights are modified as

$$W_{i,k,j} = W_{i,k,j} + \eta E_{k,j} Y_{i,j} \qquad (9.8)$$

where η is the step size of the gradient method and is sometimes called the learning rate. Therefore, the weight matrix between the output and hidden layers can be updated from the output layer to the hidden layer. The next pattern is then presented and the calculations are repeated.

The training procedures are iteratively carried out until the neural network can represent the relationships of these data within the given error tolerance. The non-parametric model is obtained by the fully connected structure and values of the weights. If well trained, the neural network can identify the system in terms of the input-output data.

The standard generalized descent method, in minimizing the objective function, is known to have several major drawbacks (Cooper, et al., 1992; Donat, et al., 1990; Leonard and Kramer, 1990). First, it adapts the weights based on the error from a single pattern representation instead of the overall error E which is the true objective function to be minimized. Secondly, the rates of convergence of this algorithm are extremely slow and the improvement per iteration falls sharply. Third drawback is that this method often leads to a local minimum.

To overcome the deficiencies of generalized descent method mentioned above, conjugated gradient method can be used to train the network. This method adapts the weights based on the overall error which is the true objective function to be minimized. Therefore it is more robust and has better convergence property than generalized descent method.

To minimize the overall error E, the conjugated gradient method updates the weights at $n+1$th iteration as

$$W^{n+1} = W^n + \alpha^n s^n \tag{9.9}$$

where W is the vector of weights and α is found by carrying out a unidimensional line minimization in the given search direction s. The new search direction at each iteration is found by

$$s^n = -g^n + \frac{\parallel g^n \parallel^2}{\parallel g^{n-1} \parallel^2} s^{n-1} \tag{9.10}$$

with $s^{(0)} = -g^{(0)}$. The elements of the steepest descent vector g are given by:
Hidden to output layer:

$$g_{i,j} = \sum_{l=1}^{p} (Y_{j,2}^{(l)} - D_j^{(l)}) Y_{j,2}^{(l)} (1 - Y_{j,2}^{(l)}) Y_{i,1}^{(l)} \tag{9.11}$$

Input to hidden layer:

$$g_{i,j} = \sum_{l=1}^{p} \sum_{m=1}^{N_2} (Y_{m,2}^{(l)} - D_m^{(l)}) Y_{m,2}^{(l)} (1 - Y_{m,2}^{(l)}) Y_{j,1}^{(l)} (1 - Y_{j,1}^{(l)}) W_{i,j,1}^{(l)} X_i^{(l)} \tag{9.12}$$

9.4   Application to Paper Machine

From an engineering point of view, neural network modeling involves unknown parameters (the weight values) and given network topology (units, layers and connections). It can be referred to as a non-parametric model that is

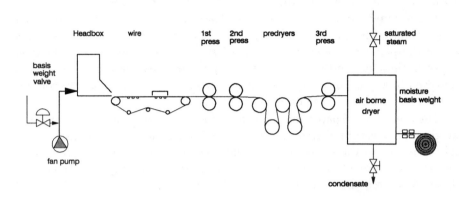

Figure 9.5: Schematic diagram of paper machine

totally different from conventional mathematical models. By this definitions, neural networks can be used to represent the input-output data and relationships of any physical systems, as long as sufficient input-output system data are collected.

The paper machine investigated is a modern fourdrinier machine producing paper board of basis weight $400g/m^2$. Figure 9.5 shows a schematic diagram of the paper machine.

The thick stock from a pulp storage is pumped into a high-level tank in which the level and the consistency of the pulp are controlled to keep in constant. The pulp from the high-level tank flows through a basis weight valve and is diluted by white water at the inlet of a pressurized headbox. The main operation objectives of the headbox are to maintain the liquid level right above the rectifier roll and adjust the total head so as to maintain the rush/drag ratio at 1.02. The liquid level is controlled by manipulating fan pump speed and the total head is controlled by manipulating the vacuum air valve opening (refer to Chapter 2). The diluted pulp in the headbox flows from the slice and onto the fourdrinier as the effect of total head.

The pulp which comes to the fourdrinier flows through the drainage devices (foils and vacuum boxes) to remove water and induce turbulence into the sheet for good formation. Then the paper sheet of dryness between 16 to 18% comes to the couch (with suction box inside) and the couch will raise that dryness to about 18 to 22%. After that the paper sheet is conveyed from the forming unit on a series of felts, rolls and open draws through the first and second presses to the dryer section. The primary functions of the press section are to remove water, consolidate the sheet, impart factorable sheet properties and promote higher wet web strength for good runnability in the dryer section.

The residual water in the sheet which leaves press section must be removed by evaporation in the dryer section. The paper first goes through predryers where the paper is heated by a series of steam-heated cylinders. And then it passes through the third press. The paper sheet leaving the third press is conveyed to an air borne dryer for final water removal. The moisture content at the end of the air borne dryer is about 10%.

The objectives of paper machine operations are to maintain the ream basis weight and moisture content at the desired values and confer desired physical properties to the paper product, meanwhile to maximize the production rate and minimize fiber material and thermal energy consumption. To achieve these objectives, the basis weight must be kept close to the lower bound of quality specifications and the moisture content to the upper bound. However, without accurate information and control of the quality variables, the operators run the machine with a conservative quality of the desired product in order to avoid off-specification product and hence deterrent to increase the production rate and reduce fiber and thermal energy consumption. Therefore, quick and accurate basis weight and moisture content prediction is crucial in paper machine operations.

For the quality measurement, on-line basis weight and moisture content sensors could be used. However, as have been emphasized in Section 9.1, paper machine includes a long series of operations. Even though the machine speed is as high as 600 feet/min, the time delay from the basis weigh valve to ream basis weight is still as large as 20 minutes, and that from the steam flow rate to moisture content is as large as 15 minutes. Therefore, even with accurate sensors for basis weight and moisture content, the neural network based prediction is still necessary for taking proactive action.

The most crucial problem for the successful application of neural network in industrial processes is the data selection and preprocessing. The industrial historical operating data are usually noisy, poorly distributed, and spotty. These issues, if not properly addressed, can prevent neural network from yielding stable models and reliable results (Stein, 1993).

The phases of data preparation for neural network training can be broadly classified into three distinct areas: data specification, in which variables of interest are identified and collected; data inspection, in which data is examined and analyzed; and data preprocessing, in which some data may be restructured or transformed to make it more useful. These steps are typically performed iteratively and in parallel with each other and the actual fitting of the model itself.

First of all, the input variables of the neural networks should be identified, that is data specification. The goal of modeling is to give the most accurate model using the fewest variables and simplest structure. Therefor, we need to identify the more useful input variables in order to reduce the problem space. A fundamental understanding of the problem domain is required for this purpose. For example, we need to determine which variables contribute variations in basis weight and moisture content; can the other variables capture the same information; is the variable significant to the quality variables; is the variable always changing or kept in constant.

The basis weight and moisture content are affected by quite a few operating conditions which include:

- Kappa number,
- thick stock consistency,
- basis weight valve setting,
- headbox consistency,
- fan pump speed,
- headbox level,
- headbox total head,
- production rate,
- wire speed,
- slice opening,
- rush/drag ratio,
- vacuums for 12 vacuum boxes,
- press loadings for 3 presses,
- vacuums for 3 presses,
- vacuums for 3 uhle boxes,
- air borne dryer steam pressure,
- air borne dryer steam temperature,
- predryer steam pressure,
- predryer steam temperature.

If we consider all these affecting factors, there will be over 40 inputs to the neural networks, which implies a huge network requiring at least thousands of training data and heavy computation burden. In order to avoid this problem, the inputs to the neural network are analyzed. Four kinds of variables will not be considered as the inputs of the neural network:

(1) The variables which have minor influence on the quality variables, such as steam temperatures;

(2) The variables which are well controlled and are thus always kept constant, such as the thick stock flow rate;

(3) The variables which might be important to quality prediction but not readily available. For example, Kappa number has effect on both basis weight and moisture content. However, the time delay from the Kappa number measurement to the pulp being consumed in paper machine is over 24 hours. Moreover, the value of Kappa number has been averaged in storage chest and machine chest. Its effect on the paper quality is hard to evaluate.

(4) The variables which are dependent each other. For example, the fan pump speed completely determines headbox consistency given thick stock consistency and flow rate.

The importance of an input variable to a quality variable can be evaluated by calculating the correlation coefficient for them. The correlation coefficient of two variables gives an indication of the strength of the relationship between the two. It measures the degree to which two variables move together. The correlation coefficient is a value from -1.0 to 1.0. A value of zero indicates no correlation, values close to 1.0 and -1.0 indicate high positive and negative correlations, respectively. Another method is analyzing trained neural networks to determine the significance of input variables, for example calculating the derivatives of trained neural networks by introducing small change in the input variable. The selection of input variables involves tradeoffs between prediction accuracy and complexity of the neural network.

After careful data specification, the final input variables for ream basis weight network are: basis weight valve setting, thick stock consistency, headbox consistency, slice opening, production rate, wire speed; the input variables for moisture content network are selected as: air borne steam pressure, production rate, wire speed, rush/drag ratio, headbox consistency, average value of vacuums of vacuum boxes; average value of press loadings.

After the inputs are identified and data are collected, we must determine whether the data need preprocessing by examining the distribution of each variable. The distribution of a variable gives information about the relative frequencies of value that a variable has taken on within a specified range. Neural networks tend to perform better when the input data is normally distributed. Histogram, which slice the range of possible variable values into equally-sized bins, is a common way for examining frequency distributions. Once the data have been collected, examined, and is ready to use, we manipulate the data in order to squeeze more information out of it. One of the common techniques for treating nonnormal data is to perform a nonlinear transform to the data, such as logarithms $ln(x)$.

In our applications, we collected the historical data for one full year. It is found that most of the data have minor deviations from the desired values. If we use such kind of data to train the neural network, the prediction from the network will always close to the given static operating condition. To solve this problem, we selected data set based on the distribution of basis weight and moisture content. For some regions of the distribution for which the historical data do not cover, on-site experiments are conducted to obtain the data. The basis weight network and moisture content network use 2000 and 2300 sets of data respectively, which are close to normal distribution, for off-line training.

As pointed out in Section 9.3, standard backpropagation based on steepest descent algorithm is very slow to converge because of the constant learning rate. In order to increase the learning speed of the neural network, a new learning rate function $\eta(t)$ is introduced

$$\eta(t) = \begin{cases} \lambda_0 & \| \Delta W_{i,j,k} \|^2 > \delta \\ \lambda + \lambda_0 e^{-\mu t} & \| \Delta W_{i,j,k} \|^2 \le \delta \end{cases} \tag{9.13}$$

where λ and λ_0 are parameters to be chosen, ($\lambda_0 > \lambda > 0$, $\mu > 0$ and $\delta > 0$). They are system dependent and usually determined from experience and knowledge on neural networks.

In the initial stage, $\eta(t)$ has a relatively larger value because λ_0 is chosen larger, $\lambda_0 > \lambda$, in order to promote a fast learning convergence. When the connection weight matrix is convergent to a certain degree ($\| \Delta W_{i,j,k} \|^2 \le \delta$), $\eta(t)$ starts decreasing until $\eta(t) = \lambda$, and λ is selected relatively small for better convergence in the final stage.

The neural network selected for predicting basis weight has 6 neurons in the input layer ($N_0=6$) with 20 neurons in the single hidden layer ($N_1=20$) and 1 neuron in the output layer ($N_2=1$). The number of data in training data set is 2000. λ_0, λ and μ are 0.5, 0.09 and 1.0, respectively.

The neural network selected for predicting moisture content has 7 neurons in the input layer ($N_0=7$) with 25 neurons in the single hidden layer ($N_1=25$) and 1 neuron in the output layer ($N_2=1$). The number of data in training data set is 2300. The selection of λ_0, λ and μ are the same as in the network for the basis weight.

We have also applied the conjugate gradient method for neural network training. It is shown that these two methods give the similar results, however, the conjugate gradient method gives better convergence.

The ability of the neural network model for fitting the trained real plant data is shown in Figures 9.6-9.7, and that for the untrained real plant data

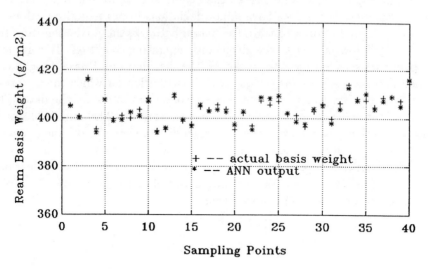

Figure 9.6: Neural network fitting the trained basis weight data

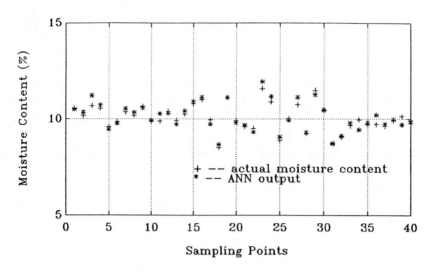

Figure 9.7: Neural network fitting the trained moisture content data

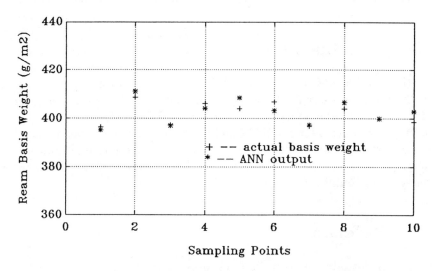

Figure 9.8: Neural network predicting basis weight

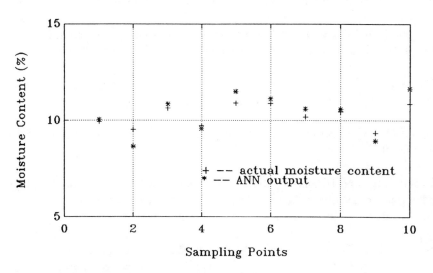

Figure 9.9: Neural network predicting moisture content

is in Figures 9.8-9.9. These results show that the neural network model can predict ream basis weight and moisture content with acceptable accuracy.

9.5   Conclusions

In this chapter, it was demonstrated that neural network can model the basis weight and moisture content for a paper machine. The neural network model outputs matched the actual quality variables with satisfactory accuracy. In addition, the neural network model had the ability to fit noisy plant data. The neural network shows promise for dealing with more complex multidimensional mapping and inherent parallel architecture should allow real-time performance even for very complex mapping.

9.6   References

Behnam B. Introduction to neural networks for intelligent control. ITT Control System Magazine 1988; 4:3-7

Bhat N. and McAvoy T. Use of neural nets for dynamic modeling and control of chemical processes systems. Computers Chem. Engng. 1990; 14:573-583

Cooper D.J., Megan L. and Hinde Jr.R.F. Comparing two neural networks for pattern based adaptive process control. AIChE J. 1992; 38:41-55

Donat J.S., Bhat N. and McAvoy T.J. Optimizing neural net based predictive control. Proc of American Control Conference, San Diego, 1990, pp 2466-2471

Gallant S.I. Connectionist expert systems. Communications of the ACM 1988; 31:152-69

Hoskins J.C. and Himmelblau D.M. Artificial neural network models of knowledge representation in chemical engineering. Computes Chem. Engng. 1988; 12:881-90

Hudson D.L., Cohen M.E. and Anderson M.F. Use of neural network techniques in a medical expert system. International Journal of Intelligent System 1990; 6:213-23

Kim H.C., Shen X., Rao M., McIntosh A. and Mahalec V. Refinery product volatility prediction using neural network. Proc 42nd Canadian Chemical Engineering Conference, Toronto, Canada, 1992, pp 243-245

Minsky M. and Papert S. Perceptrons. Cambridge, MA, MIT Press, 1969

Lenoard J. and Kramer M.A. Improvement of training neural networks. Computers Chem. Engng. 1990; 14:337-341

Rumelhart D.E., Hinton G.E. and Williams R.J. Learning internal representations by error propagation. In: Parallel Distributed Processing, Vol 1. MIT Press, Cambridge, 1986, pp 318-62

Stein R. Selecting data for neural networks. AI Expert 1993; 2:42-47

Venkat V. and King C. A neural network methodology for process fault diagnosis. AIChE J. 1989; 35:1993-2002

Wasserman P.D. Neural computing: theory and practice. Van Nostrand Reinhold, New York, 1989

CHAPTER 10
IOMCS FOR PULP AND PAPER PROCESSES

The previous nine chapters present a number of control algorithms and their applications to unit operation of paper machines. The growing complexity of industrial processes and the need for higher efficiency, great flexibility, better product quality, lower cost and environment protection have changed the face of industrial practice. Mill wide information management, decision-making automation and operation support have become very crucial for the modern pulp and paper company to stay competitive internationally. Artificial intelligence can play an important role in attaining the above goal. Considering that most of the nowadays modern pulp and paper mills have successfully installed distributed computer control systems (DCS) and information management systems, it is of very significant economic benefit to improve the existing systems by adding "intelligence" and enhancing functionality. This chapter presents an intelligent on-line monitoring and control system (IOMCS) for pulp and paper processes. IOMCS is a real-time intelligent system which links with DCS and a mill wide information system. It takes advantage from the DCS value-added data in the information management system. The system fulfills functions such as monitoring the process for abnormal situations; advising evasive and corrective operation actions to operators; pulp quality prediction; and operation optimization. Unlike in the previous chapters, we will not limit our attention to paper machines, but to whole pulp and paper processes. The techniques proposed in this chapter are applicable to paper machines.

265

10.1   Introduction

Pulp and paper production is a resource and energy intensive industry. Therefore, the rewards associated with the successful control and optimization of pulp and paper processes are very large indeed (Matson, 1989).

Pulp and paper production consists of a complex and long sequence of operations starting from a wood chipping plant, ending at a paper machine. Knowledge used in the operation includes complex technologies from different areas. Operators have to monitor an increasingly large amount of raw data and supplemental information during a potentially critical situation. A process operator, even an expert, is unable to possess all kinds of operation knowledge. He may be capable of dealing with routine operations. However, when facing operation mode changes and emergency situations, the operators may find it difficult to contribute a quick and effective solution.

The development of industrial process automation may be divided into four stages (Rao, 1991): (1) labor-intensive stage; (2) equipment-intensive stage; (3) information-intensive stage; and (4) knowledge-intensive stage. A lot of modern pulp and paper mills have been in the third stage, in which a distributed control system (DCS) and a mill wide information system have often been successfully installed. For example, most of pulp and paper mills in Alberta, Canada have installed DCS and a mill wide information system called MOPS. MOPS has such functions as display handling, trend handling, material tracking, statistical process control, cost reports, etc. These functions help operators as well as managers to check the status of the mill quickly, make decisions efficiently, access the operating conditions for new production of repeated grades. The benefit from implementing MOPS has been very significant (Frith, et al., 1992; Henriksson, et al., 1992). Using MOPS, however, the mill operation still relies on individual operators' experience. Even though consistent and thorough operator training, individual start up experience has made operators react differently to similar situations. To deal with these problems, Intelligence Engineering Laboratory at the University of Alberta are working together with several pulp and paper companies to develop an intelligent on-line monitoring and control system (IOMCS) (Xia, et al., 1993).

Up to now, the research on pulp and paper process automation is mainly on model based advanced control techniques (Dumont, 1990; Xia and Rao, 1992). The operation of pulp and paper process represents a knowledge-intensive task. Knowledge-based expert systems are more suitable than the model-based techniques to deal with such kind of problems that require considerable expertise (Dvorak and Kuipers, 1991). They are superior alternatives to numerical computation since they can capture and utilize more broadly-based sources of knowledge. The planning, scheduling, monitoring, analysis and control of process operations can benefit from improved knowledge-representation schemes and advanced reasoning control strategy (Stephanopoulos, 1990). There are a number of researches on knowledge-based systems on pulp and paper processes, such as intelligent operation support system for batch digester (Rao and Corbin, 1993), diagnostic expert system for solving pitch problems (Kowalske, 1991), decision support system for pulp blending strategy (Dane and Harvey, 1992), and expert system for the on-line monitoring of waster water treatment process (Lapointe, et al., 1989). There are also some researches related to intelligent monitoring systems for some industrial processes other than pulp and paper process. Murdock and Hayes-Roth (1991) proposed an architecture to augment classical control method with artificial intelligence method to monitor semiconductor manufacturing. Macchietto and his coworkers (1989) presented a system to solve on-line optimization problems utilizing the features of a general purpose flow sheet simulation. These systems function well to perform single task for some specific processes. However, they do not consider the integration with the existing systems in mills, and cannot provide the integration of different problem solving techniques.

IOMCS is radically different from the previous model based numerical computation systems and rule based expert systems. It is a real-time intelligent system that links with DCS through MOPS and takes advantage from the DCS value-added data in MOPS. It integrates different problem solving techniques to solve the complex engineering problems. Its main functions are: monitoring process variables and advising evasive or corrective actions to operators to recover the production from undesirable situations; predicting changes in the final quality of pulp; decision support for upset conditions. IOMCS also has a hypermedia information module that can be used by operators for on-line help as well as for training. IOMCS is an artificial intelligence application for knowledge-intensive automation.

10.2 System Design and Implementation

10.2.1 Functional specifications

IOMCS is an integrated distributed intelligent system that fulfills on-line monitoring and control tasks. The system aids process operation in two ways: (1) it generates optimal operating conditions in normal situations and corrective actions in abnormal situations and interfaces with MOPS and DCS to influence the pulp processes; (2) it makes feasible operation recommendations

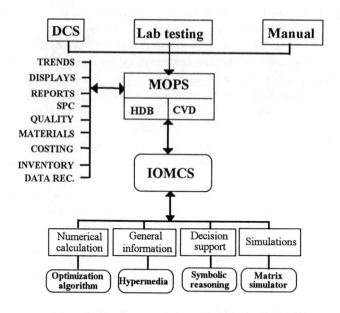

Figure 10.1: Interconnection of IOMCS with MOPS

to operators who make final decision and take actions.

One of the main issues considered in the design of IOMCS is protecting companies' previous investment. The objective of developing IOMCS is to make full use of the available computer facilities in mills and to enhance pulp production from information management to decision automation (Xia, et al., 1993). The intelligent system does not intend to replace the existing DCS and information management system, but to add "intelligence" to and enhance the functionality of the existing systems. IOMCS is a computer software system implemented based on the existing MOPS system. It acquires data from MOPS CVD (current value data base) and HDB (historical data base) and uses the value-added data for automatic decision making. The companies need to invest very little in hardware to install IOMCS. Figure 10.1 presents the interconnection of IOMCS with MOPS and DCS. Not only is IOMCS an intelligent system for on-line monitoring and control of pulp and paper production, but also provides an intelligent system building tool embedded in MOPS.

Another main issue considered is keeping operators in decision making and control loops. IOMCS does not intend to replace human operators in decision making, but to help them by providing highly concentrated and understandable information from vast amount of data. The multimedia operator interface can provide information and explanation with natural language, numerical values, and graphical display. Thus, operators get "closer" to the production processes and perform better operation.

In general, IOMCS provides the following features:

- standardization of operators' changes,
- more timely and knowledgeable decisions,
- identification of complex process situations that have the potential of causing process upsets,
- automatically learning of optimal operating conditions for different product grades, and
- having an open structure to implement the local systems independently and then integrate them into a mill-wide system.

Through a multimedia operator interface, operators are able to

- view an acknowledge situations and advisories,
- view "how & why" type information and access to further process information if required,
- scroll through old messages and sort by process area,
- provide an interface for process engineers to program the knowledge base.

10.2.2 System design

In order to meet the requirements of pulp production, IOMCS possesses the following additional desired properties:

Process generality: IOMCS is general and flexible in nature, and not process specific. The methodology applied enables the system to easily accommodate the changes in the plant configurations.

Application-specific shells: The use of IOMCS does not require much knowledge in computer science or artificial intelligence. Users can implement IOMCS on a specific process just by making revision in process-dependent knowledge bases.

Open architecture: Developing a mill-wide IOMCS in one phase is very difficult. IOMCS has an open structure such that different applications can be implemented step by step, independently. In our project, we develop the system for the bleach plant and final mill quality as the first stage. The next system will be for recovery boiler, then for effluent treatment, and so on. Finally, all these local systems will be integrated into a mill-wide system.

Multiple knowledge representations: The knowledge required to represent the declarative and procedural information in IOMCS is from many different fields, which includes:

- processes,
- optimization methods,
- control technologies,
- fault diagnosis technology,
- process operating conditions, and
- product specification.

In order to represent all the above different knowledge, IOMCS facilitates multiple knowledge representation techniques, such as facts, rules, numerical models, neural networks and graphics simulation.

The performance criteria considered in IOMCS include:

- to increase productivity,
- to improve product quality,
- to reduce cost (raw materials, chemicals and energy),
- to reduce environment load, and
- to improve customer service.

Pulp and paper production is a large-scale manufacturing process. IOMCS for pulp and paper processes deals with combinatorially large problems. When solving these problems for large-scale processes, the result is a prohibitively costly and slow solution. To reduce the problem solving cost, we need to separate the pulp and paper mill into a number of units, such as refining, bleaching and effluent treatment. In this way, we break up a large problem into a number of simpler problems. Distributed intelligent systems and cooperative expert systems are the technology and methodology of breaking up a problem into multiple agents, and distribute the reasoning and computation at different levels of the automation pyramid. The topology of interconnection relationships between the processes and problems should be specifically modeled. In this way, IOMCS can perform optimal analysis across process components and operating environments, and to allow for flexible manufacture.

From the discussion above, IOMCS integrates and cooperates, both hierarchically (different level of automation pyramid) and latitudinally (process decomposition), information, management and process. By using an open structure, it can be extended from a single process step, to mill-wide applications. The methodology for this kind of integration and cooperation is key to the next-generation of industrial application intelligent environment.

Three-layer architecture

Based on considerations above, IOMCS can be characterized as a domain-dependent solution shell that is a flexible and reconfigurable application for

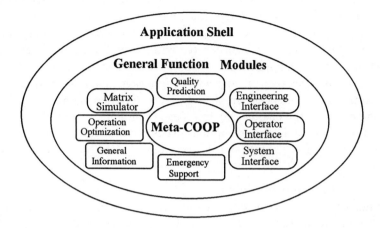

Figure 10.2: Three layer architecture for IOMCS

solving the control problems in pulp mills. It is designed using general function modules. Users can select functions from the available function modules
and implement them according to the equipment and process layout. Modu-
lar technology makes IOMCS more flexible and easier to implement (Soucek,
1991). A number of function modules are identified to be important for pulp
processes:

- matrix simulator,
- quality prediction,
- operation optimization,
- emergency support,
- general information,
- operation interface,
- engineering interface, and
- system interface.

The system architecture that has the above characteristics is of three layers,
as depicted in Figure 10.2.

The first layer that is the core of IOMCS is an integrated distributed intelligent system shell, Meta-COOP. It provides an inference engine, and facilitates
knowledge representations, acquisition, editing and compiling.

The second layer provides general function modules for on-line monitoring
and control. It is a medium between the system core (Meta-COOP) and
specific applications. In this layer, a number of modules are developed by

implementing the general monitoring and control technologies into computer software supported by Meta-COOP. Each module fulfills a typical monitoring and control function that is necessary in pulp mill operations. Every module can be run independently or incorporated with several other modules.

The third layer is an application-specific shell for users to implement IOMCS for some specific processes. It applies the function modules provided in the second layer. The users can simply select the function modules, and enter knowledge by using the engineering interface. Therefore, not much experience in computer science and artificial intelligence is required for implementing specific applications. The process and process control engineers in pulp mills are able to do it.

The general function modules

The function modules corporate to fulfill the on-line monitoring and control tasks. The **General Information** module provides general mill information including text-like and graphical information by applying intelligent hypermedia technology. Intelligent hypermedia is a promising technology to organize heavily referenced information with changing variables in a more efficient way. It links a keyword or phase to text or graphics that further explains the original keyword of phases. The intelligent hypermedia system has reasoning and decision making capabilities. Users can define variables and their relationship in text and graphics by using rules and procedures. In the General Information module, all the mill information is organized into discrete blocks or nodes of information, called topics. Links are then used to join these topics to one another to form the document as a whole. The information includes:

- main process equipment in the mill,
- brief introduction of the process,
- architecture and functions of DCS, MOPS and IOMCS,
- hazard alerts,
- grade recipes,
- mode of operation for all grades,
- procedures for startup and shutdown operation, and
- procedures for emergency operation.

Operators can easily find subjects of interest by going through a menu system or searching for keywords and topics in text and graphics. Figures 10.3 and 10.4 show some information about wood structure and fiber morphology included in the general information module. By clicking the linking word, "see diagram", in the text (Figure 10.3), a picture of the cross section of tree is accessed immediately (Figure 10.4). This module can serve as a training tool

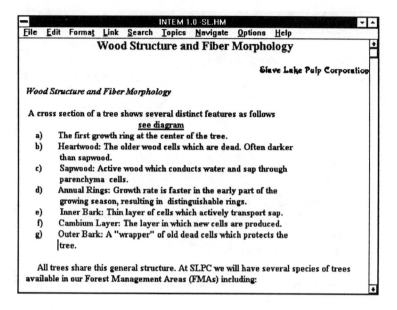

Figure 10.3: Text-like information in the general information module

for new operators to become familiar with operations. It can also be used for on-line help.

The **Matrix Simulator** is a process simulation package and graphical tool built in IOMCS. In the Matrix Simulator, process variables are divided into two parts: action variables and result variables. Action variables affect the product quality and can be changed to rectify the process conditions. Result variables that are affected by the action variables are either product qualities or those used to evaluate the performance of the production processes. The matrix simulator provides the relationship between result variables and action variables. Unlike conventional simulation packages, the matrix simulator is built based on the expert system building tool, Meta-COOP.

The **Quality Prediction** module applies the matrix simulator to predict quality changes. The accurate measurement of some important quality variables such as freeness and brightness is not instantly available. This module fulfills tracking task to advanced model's state in step (forward reasoning in the matrix simulator) with observations from the physical system. The laboratory test results are used to correct the prediction. Because of the complexity of pulp processes, it is impossible to obtain complete process models using conventional methods. On the other hand, there are a lot of historical data

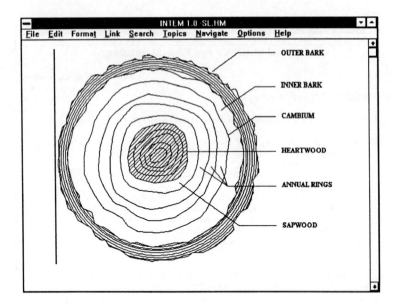

Figure 10.4: Graphical information in the general information module

available in the pulp mills. Neural network technology is thus applied as a complement of conventional modeling techniques for quality prediction.

The **Operation Optimization** module applies backward reasoning in the matrix simulator and the real-time value-added information in MOPS to learn the optimal operation conditions that result in the best product quality and lowest costs. An optimization package called SWIFT is developed using FORTRAN language. SWIFT can perform numerical optimization with equality and inequality constraints. The optimization criterion is to produce pulp economically, which implies: making the best use of raw materials, controlling waste, limiting fiber losses and providing better product quality.

The **Emergency Support** module monitors undesired conditions and diagnoses the original source of the process upsets. The sequence of corrective actions to recover the process from the undesired situations is automatically generated after an upset is detected.

IOMCS has many different features. In the remainder of this chapter, we introduce in more detail about the matrix simulator, the system interface and emergence support system.

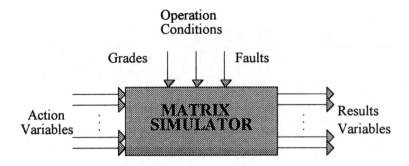

Figure 10.5: Illustration of matrix simulator

10.2.3 Matrix simulator design

Matrix simulator was originally developed for the PDP-11 operating system. The code was written in FORTRAN language by SCA (Sweden's second largest pulp and paper company). This product was ported to VAX/VMS environment and was using "home"-made database. The user interface in this system was a PC-based software product called Eyes-Cream. However, this product was not compatible with MOPS and it was an off-line numerical based simulation tool. In our project, we used the object oriented features of Meta-COOP to develop a knowledge based matrix simulator.

In the design of the matrix simulator, the first step is to describe the problems in terms of result and action variables. Matrix simulator is actually a functional and structural model base of the production process. In order to improve the flexibility of the matrix simulator, the global pulp and paper process is divided into numerous units. The action and result variables for each unit are defined. The global matrix simulator integrates the relationships among all units according to material and signal flow. The next step is to define the normal operating conditions of production. They are defined using grades and production range. A grade defines the final quality of a particular product. The production range is used to incorporate different relationship among variables when the process operates at different speed and load. For each grade, the target, maximum and minimum limits are also defined. The matrix simulator receives process data from MOPS and identifies the normal operating conditions (product grades and production ranges). According to the normal operating conditions, one relationship between the action variables and result variables is selected. Figure 10.5 gives an illustration of the matrix simulator.

The matrix simulator displays using an object oriented drawing tool, PI-

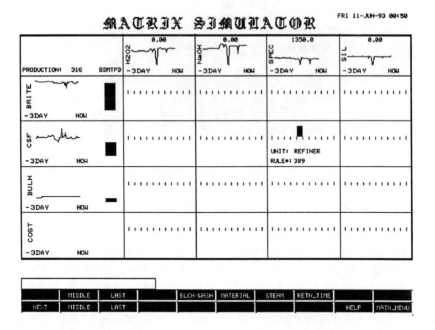

Figure 10.6: Matrix simulator's user display

CAD, which is a function of MOPS and is PC based. It displays in PCs using MOPS' operator station interface program, EDE/2. In order to compile this display in VAX environment using PIC (MOPS VAX based picture compiler), we must allocate a unique address in MOPS database for each of the dynamic points of the displays using MOPS Point Configure Editor (PCE).

The operating display of the matrix simulator consists of many squares as shown in Figure 10.6. The bargraphs display the relationship between action and result variables. If the bargraph is on the positive side of the zero line, an increase in the action variable will cause an increase in the result variable. If the bargraph is negative, an increase in the action variable will move the result variable towards zero line. The other boxes, besides action and result variables, display trend curves, target, minimum limit and maximum limit. The other important information on the matrix simulator display is:

- Grade information displays the target grade of the final product.
- Grade recipe ID informs operators which operating conditions are currently selected. For each result variable we also have three bargraphs:
- The current value bargraph displays the current situation. The target,

maximum and minimum values are also displayed along this graph.

- The simulated bargraph displays the effect of an action variable change on the result variable.
- The dynamics bargraph displays the predicted value of the result variable when the next sample is taken from the process. Hence, it takes into account the dynamics and retention time of the process.

For the action variables, we display two values and their respective trend curves. One of these values is the actual current value of the action variable and the other is the simulation value a user would enter.

The operator can introduce changes in action variables (perform an operation action) and observe the corresponding changes in result variables from the graphical displays. Other function modules can call the matrix simulator for different purposes.

10.2.4   System interface: IOMCS communication with MOPS

MOPS collects data from many different sources and stores the data in its database. Sources include distributed control systems, programmable controllers, supervisory computers, data processing systems, laboratory instruments, computer terminals and others. Hence, MOPS can play a pivotal role in integrating the mill's process control systems with its computer-based manufacturing applications such as production management, engineering and maintenance. Therefore, an efficient communication link between MOPS and IOMCS is very important. By interfacing with the MOPS database, IOMCS system would have access to more than 2000 points per minute of data collected over the last three years at its disposal. To reduce the data traffic in the mill networks, matrix simulator is so designed that it can be used directly by operators at their stations or other programs within VAX environment.

MOPS is installed in VAX but displayed in PCs. Besides, the intelligent hypermedia system applied to implement the General Information module is also installed in PC's. Therefore, IOMCS must communicate with both VAX and PCs.

IOMCS communication within VAX: This is achieved by adding two function routines to IOMCS. The first function is called *"getdata"*, which is for getting the requested data from the MOPS database. The second function is called *"putdata"*, which is for putting the requested data to the MOPS database.

IOMCS communication with PCs: These are the steps taken to embed the IOMCS into MOPS:

Step 1 IOMCS is activated in every monitoring time interval or on the request of operators. Then *matdsp* routine gets activated.

Step 2 MOPS gives the static part of the user displays and returns the control to IOMCS by executing *usinit* function.

Step 3 IOMCS gets all the required information from MOPS database with *usgepn* routine.

Step 4 IOMCS does its reasoning and decision making, then uses *usput* to put the reasoning results in MOPS database and informs MOPS with *uscomp* routine. The results include the dynamic part of the matrix simulator, explanations about the current production status, operation suggestions to operators, etc.

Step 5 MOPS sends these results to the MOPS' operator station interface program, EDE/2, on PCs.

Step 6 EDE/2 sends some of the results it received to Clipboard; the intelligent hypermedia system gets these results from Clipboard and display specific topic according to the received information.

Step 7 IOMCS sleeps until next monitoring time comes or a new user requests using *usgtcm* command, and then goes to step 1.

10.2.5   Emergency support system

We focus in this section on real-time fault diagnostics and emergence support ability of IOMCS for a bleach plant.

The process overview of the bleach plant is given in Figure 10.7. The product quality is maintained by assuring that process variables fluctuate within permissible ranges. If operating conditions go beyond these limits, the product quality is in jeopardy. Diagnosis of process faults (process upsets) can be very time consuming. This makes it difficult for operators to always take the correct action. In a mill environment where time is critical and many things are going on at the same time, hesitation or improper action could lead to serious implications. Automated fault diagnosis can offer a unique aid to operators in detecting, isolating and handling process upsets. Besides, fault diagnosis and emergency handling decision making include lots of experts' private knowledge and heuristic knowledge. Artificial intelligence is an very promising technique to be applied here.

Fault diagnostic systems have been a main research field for artificial intelligence (Qian, 1990; Kramer and Leonard, 1990; Yu and Lee, 1991). Numer-

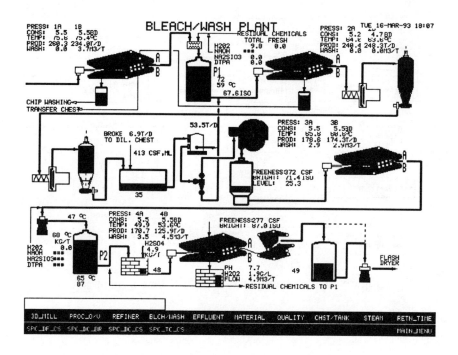

Figure 10.7: Overview of bleaching plant

ous papers report applications of expert system to alarm monitoring, failure detection and diagnosis. Several techniques can be used to perform fault analysis such as fault tree analysis (Shafaghi et al., 1984), signed directed graphs (Kramer and Palowitch, 1987) and artificial neural networks (Hoskins et al., 1991).

Bleach plant affects almost all the pulp quality parameters. The quality variables considered include

- brightness (ISO, %)
- freeness (CSF, ml)
- bulk (cm^3/g)
- breaking Length (m)
- tear Index (mNm^2/kg)
- burst Index (MPam^2/kg)
- scatt Coeff, (m^2/kg)
- opacity (%)

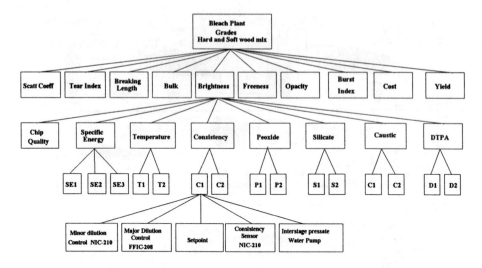

Figure 10.8: Variable relation tree of bleach plant

- cost ($/t)
- yield (t/t)

These quality variables are dependent on the consumption of caustic (kg/t), peroxide (kg/t), silicate and DTPA, and the pulp consistency (%), specific energy (kwhr/t) and wood species and chip quality. The relation among the quality variables and the affecting variables can be represented by Figure 10.8.

The functions of emergency support system include monitoring operation performance, detecting undesirable conditions and recommending emergency handling procedures (corrective action) to the operators.

The real-time emergency support system for a bleach plant faces some additional requirements:

1. Multiple faults: More than one physical components may be the source of plant failure. Multiple independent faults can be diagnosed exactly like single faults because their symptoms propagate through non-intersecting systems. Dependent faults are difficult to diagnose, because there is a common mode of failure.

2. Propagated faults: The processes and components in a bleaching plant interact with each other. A single component failure may trigger a sequences of faults. The system should be able to locate the original fault.

Moreover, when some sensors degrade in performance, the system must still continue to generate an acceptable diagnosis.

3. Closed loop compensation: Since the bleach plant is controlled by DCS, a failure in any component can either be magnified or compensated due to interactions within the feedback loops. The system should be capable of identifying precisely the source of failure even with the feedback compensation.

4. Multiple operating conditions: The different grades of products are produced according to customers' requirements, which results in multiple operating conditions. For different operating conditions, the rules and mechanism for fault diagnosis will be different. The system should be able to learn the operating conditions from the process measurement and automatically adjust the fault diagnosis mechanism.

5. Timely response: The system must complete its reasoning quickly. Any delay in decision may result in production loss, off grade or equipment damage.

To meet the above requirements, in the fault diagnostic system we use the following fault modeling techniques:

- structure and function modeling;
- fault propagation modeling; and
- hierarchical decomposition.

These modeling techniques are carried out according to the variable relation tree depicted in Figure 10.8.

Structure and function modeling

A failure can be defined as a violation of expected plant behavior or of certain function constraints. Therefore, to detect and locate the failed component, we must model the plant based on both function and structure.

All the functional and structural models are built in the matrix simulator. The inputs of the matrix simulator are action variable, which are the process operating variables affecting product quality. The outputs of the package are called result variables, which are product quality parameters. Because of the complexity of the bleach plant, it is impossible to obtain complete functional and structural models only by conventional techniques. Neural network technology is applied here as a complement of conventional modeling technique. For this purpose, we modify and apply a neural network software which was originally developed for refinery plant (Kim, et al., 1992). The matrix simulator consists of Meta-COOP based simulation models and artificial neural networks. The emergency support system communicates with the matrix simulator through an interface to obtain the input- output relationship. The matrix simulator is illustrated in Figure 10.5.

Fault propagation modeling and criticality analysis

A fault propagation is a simple and powerful method of expressing the dynamics of fault evolution. A failure in the model is connected only to its immediate effects, which in turn propagates to other effects. Using a forward chaining search from a component, the system can simulate a failure of that component in terms of a time sequence of symptoms. After detecting an initial fault, the system will predict the impending faults. According to the analysis of criticality of these faults (occurred and impending) to the product quality, production benefits, and safety, the system will make decisions and suggest the operators to handle the emergency event to avoid or minimize production loss. Using a backward chaining search from the symptoms, we can locate fault sources.

The fault propagation model is described by a fault propagation graph, a casual relationship on a set of failure modes. Two kinds of causal links are possible: Or and And. An Or casual link has one predecessor failure mode and one or several successor failure modes, and implies that the predecessor failure mode might propagate, thereby creating the successor failure mode. An And causal link has more than one predecessor failure mode and one or several successor failure modes, and implies that only the presence of all predecessor failure modes can propagate and cause the successor failure model. Fault propagation model provides the heuristic relation among the fault symptoms. The quantitative results such as the propagation intervals and gains come from the matrix simulator. Figure 10.9 shows part of fault propagation model for bleach plant.

Hierarchical decomposition

Since real-time based system deal with time critical situations, the efficiency of fault diagnosis is a key issue to be considered in the system design. Bleaching is a complicated process. Considering that we are going to develop a mill-wide IOMCS, the process is even more complex. The fault-diagnosis will be very ineffective if we treat the whole plant as a single model. This kind of single model is difficult to develop because of the complexity of the process. To solve this problem, we develop a hierarchical structural model for the bleach plant. Each node in this hierarchy is a process, an abstract functional entity that consumes and produces material, energy and information. The function

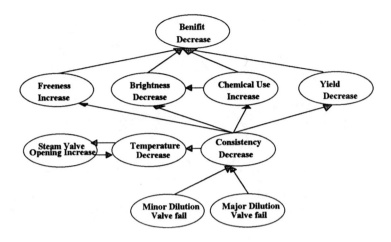

Figure 10.9: Fault propagation graph (part) for bleach plant

model for each node is developed independently.

A bleach plant is very complicated. Figure 10.10 shows only a simplified hierarchical structure model.

Since each node in the above hierarchy is a component, we must develop a normal behavior model for each nodes and connect them into a whole process model. This technique can also provide us the ability for system extendibility. For example, we first develop the IOMCS system for the bleach plant. In developing the mill-wide IOMCS system, the bleach plant serves as one node of the whole pulp mill structure model.

Using the above decomposed hierarchical model, when one or more processes violate their functional constraints due to the presence of faults, the system will give alarms to indicate the presence of failure modes. The alarm can indicate the presence of failure processes at different levels of hierarchy. The system determines how detail its diagnosis result will be given according to time/resolution trade-offs and the information available. If there are not enough on-line measured information or not enough computation time, the system will stop the diagnosis search at the higher level of the granularity. More detail (lower level) diagnosis will be conducted on the request of operators or by providing more information through multimedia operator interfaces. The information is provided to system using the dialogue between system and operator. The dialogue consists of questions and choice of answers. It should be emphasized that after the system locates a fault source at a particular node

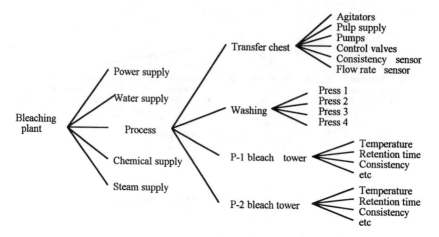

Figure 10.10: Process decomposition of bleach plant

(process) in the hierarchy, the search in the subhierarchy rooted at that node is needed only. We do not need to search other subhierarchies at the same level. That is the main reason why we can improve the searching efficiency by using the hierarchical decomposition of function and structure model.

Fault reasoning technique
   The on-line monitoring function is carried out in the following six steps:

Step 1   monitoring the process for undesirable conditions (fault detection),
Step 2   identifying the initial cause of the situations (fault isolation),
Step 3   analyzing the criticality of the situation (fault analysis),
Step 4   deciding the corrective actions to be taken (fault handling),
Step 5   issuing the corrective commands to MOPS and DCS, and
Step 6   report the alarm, cause and criticality of the situation and the suggested actions to operators.

   Fault detection detects the presence of undesired conditions or process upsets. Thus it needs only to monitor a number of most important process quality variables and some other variables which strongly affect quality. There are two cases to be considered:
   (1) Single upset: We monitor those variables whether they are within the specified upper and lower limits for different grades. If all variables are within the limits, there exists no process upset. Otherwise there are some upset. For example, if the consistency of the pulp pumped to the Twin Wire Presses is

higher than 6% or lower than 4%, one process upset presents. vspace0.0in

(2) Combinatorial upset: There is no single variable out of the limits, but the combinatorial effect of some variables will result in quality degradation from specifications. We define such case as combinatorial upset. For example, even though chemicals, temperature and consistency are all within the prespecified limits, the combined effect may result in too low or too high brightness. To decide the combinatorial upsets, the models in the matrix simulator will be used. The models (which eliminate the retention time attached to them) predict the effect of the current operating conditions on product quality and indicate whether or not the quality will be out of specifications.

When the presence of fault is detected, the fault isolation function will be carried out. For isolation purpose, both the function models (attached with retention time) implemented in the matrix simulator and the fault propagation models will be used. The tracking task advances the matrix simulator model's state in step with observations from the physical system. The propagation models serve as qualitative knowledge to determine which function model in the matrix simulator will be evaluated and call that model. When observations from the physical system disagree with predictions from the model, the fault hypothesis is generated. If all the symptoms for the fault are proved, the hypothesis is proved, which indicates that the initial fault has been isolated. After identifying a fault, the diagnosis task injects it into the current model of the matrix simulator so that the predictions will continue to match actual observations. The faulty component identification algorithm is responsible for identifying a fault source component in the context of a process' fault model. To refine the diagnostic results, the interlevel migration process migrates the identification algorithm to lower levels of the fault model hierarchy.

Fault analysis and emergency handling will use forward chaining search of fault propagation models for qualitative knowledge and the matrix simulator for quantitative knowledge. It predicts the possible events that will come from the initial fault, and applies the matrix simulator to provide the quantitative results. Each process upset is attached with a parameter, namely criticality factor, to represent the influence on the production. The decision making system decides whether the process is recoverable or not. If yes, a sequence of corrective actions will be issued to the previous operating mode. Otherwise, the process with the upset will be shut down.

An optimization package called SWIFT is developed using FORTRAN language. SWIFT performs numerical optimization with equality and inequality constraints.

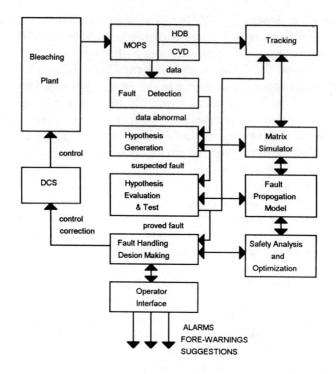

Figure 10.11: Architecture of emergency support system

We construct a benefit criteria function based on the above consideration, and consider the quality and equipment protection as constraints.

There are basically three different kinds of process upsets according to their criticality to pulp production.

(1) Tolerable upset: The upset is minor. No corrective action is required.

(2) Recoverable upset: Corrective actions are required in order for the process to run properly. The performance of the process may degrade.

(3) Unrecoverable upset: The upset is not recoverable.

The principle of the fault diagnostic technique can be represented by Figure 10.11.

10.2.6   System Implementation

We apply an object-oriented intelligent system building tool, Meta-COOP, which is developed in the Intelligence Engineering Laboratory at the University of Alberta, to implement IOMCS. The distinct characteristics of Meta-COOP make the application of the above techniques possible.

Meta-COOP is a hybrid system which permits the combination of a number of problem solving techniques. These techniques include the use of frames, rule, and a powerful programming language, C++.

Meta-COOP distributes its knowledge into a number of knowledge bases. Each knowledge base is basic object within the Meta-COOP environment called a unit. Each unit has an arbitrary number of slots, in which the attributes of the unit are described. Each slot represents one attribute of the unit and has several facets, in which the attributes are specified in more detail. There are two types of units: class units and member units. Member units describe individual objects; class units group several objects with the common attributes into a single class. Therefore, member units are "instances" of the class units. Unlike member units, class units can be defined as a superclass of subclass of another unit. The knowledge units are organized in a hierarchy with inheritance properties. With this characteristics, we can apply process hierarchical decomposition techniques to reduce a complex engineering problems to a number of less complex problems.

As a hybrid system, Meta-COOP allows the integration of various knowledge representations and inference methods, such as frame-based, rule-based and method-based, external procedures written in any other language, and internal subroutines written in C.

The intelligent system can call external executable programs, such as NEURAL, the neural networks model, and SWIFT, the optimization package.

The operator interface consists of MOPS' operator station interface, EDE/2, and the General Information module. EDE/2 gives illustrative and graphical information, such as trend curves of important operating conditions and quality variables. It also gives a brief alarm of process upsets. The detail reports of process upsets are given by intelligent hypermedia system, i.e., the General Information module. When a process upset presents, the Emergency Support module detects and isolates the event and send a message (with attached values of important process variables) to the intelligent hypermedia system to activate one specific topic (the report of the event) in the General Information module. Figure 10.12 is a example of the report. The advantage to do so is that it is extremely convenient to define a large number of emergency reports using the intelligent hypermedia. And operators can access to more detail information in the General Information module if necessary. Figure 10.13 is the Engineering Interface for mill engineers to debug the knowledge bases.

The new optimal operating conditions obtained from the optimization package SWIFT are also suggested to operators as:

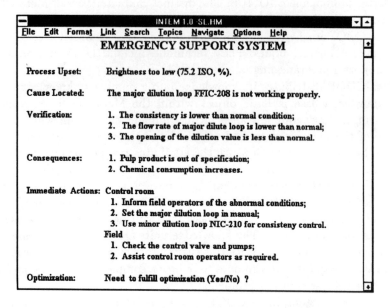

Figure 10.12: Emergency report to operators

The optimal conditions suggested are:
The consistency setting: 5.11435%
The temperature setting: 51.8449 C
The peroxide setting: 28.6956 kg/t
The silicate setting: 28.8601 kg/t
The caustic setting: 49.9063 kg/t

10.3 Conclusions

Intelligent on-line monitoring and control system for pulp and paper processes is an important and challenging research field. The presented IOMCS provides powerful operation support and on-line advice and control functions. The design of IOMCS is flexible and allows for changes in process equipment, product grades, or operating philosophy. IOMCS can also serve as a universal configuration to develop high performance intelligent systems for many complicated engineering applications.

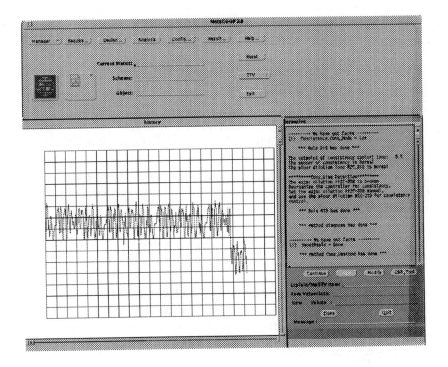

Figure 10.13: Engineering interface

10.4   References

Dvorak D. and Kuipers B. Process Monitoring and Diagnosis. IEEE Expert 1991; 6:67-74

Dumont G.A. Control techniques in the pulp and paper industry. In: Control and Dynamic Systems 1990; 37:65-113

Dyne B. and Harvey M. Decision support system for pulp blending strategy. Presented at Pulp and Paper Expert Systems Workshop, Pointe Claire, Quebec, 1992.

Frith M.D. and Henriksson C. Integration of distributed control and mill wide information systems at the Slave Lake Pulp Corporation plant. Presented at CPPA Spring Conference, Jasper, Canada, 1992

Gesser R. Building a Hypertext System. Dr. Dobb's Journal 1990; 165:22-33

Harris C.A., Sprentz P., Hall M. and Meech J.A. How expert systems can improve productivity in the mill. Pulp & Paper Canada 1990; 91(11):29-34

Henriksson C., Smith W., Danielson K. and Olofsson J. Value-added information– a high payback investiment for the mill. Proc Tappi Process Control Conference, Atlanta, Georgia, USA, 1992, pp 5-19

Hobbs G.C. and Abbot J. Peroxide bleaching reaction under alkaline and acidic conditions. Journal of Wood Chemistry and Technology 1991; 11:329-347

Holloway L.E., Paul C.J., Strosnider J.K. and Krogh B.H. Integration of behavioural fault-detection models and an intelligent reactive scheduler. Proc of the 1991 IEEE International Symposium on Intelligent Control, Arlington, Virginia, USA, 1991, pp 134-139

Hoskins J.C., Kaliyar K.M. and Himmelblau D.M. Fault diagnosis in complex chemical plants using artificial neural networks. AIChE J. 1991; 37:137-146

Kim H., Shen X. and Rao M., McIntosh A. and Mahalec V. Refinery product volatility prediction using neural network. Proc 42nd Canadian Chemical Engineering Conference, Toronto, Ontario, Canada, 1992, pp 243-244

Kitzmiller C.T. and Kowalik J.S. Symbolic and numerical computing in knowledge-based systems. In: Kowalik J.S. (editor) Coupling Symbolic and Numerical Computing in Expert Systems, Amsterdam, New York, Oxford, Tokyo, 1985

Kowalski A. Diagnostic expert system for solving pitch problems. Presented at Pulp and Paper Expert Systems Workshop, Pointe Claire, Quebec, 1991

Kramer M. and Palowitch B. A rule-based approach to fault diagnosis using the signed directed graph. AIChE J. 1987; 33:1067-1078

Kramer M.A. and Leonard J.A. Diagnosis using backpropagation neural networks–analysis and criticism. Computers Chem. Engng. 1990; 12:1323-1338

Lapointe J., Marcos B., Veillette M. and Laflamme G. BIOEXPERT–an expert system for wastewater treatment process diagnosis. Computers chem. Engng 1989; 13:619-630

Macchietto S., Stuart G., Perris T.A. and Dissinger G.R. Monitoring and on-line optimization of process using speedup. Computers Chem. Engng. 1989; 13:571-76

Matson W. The outlook for the Canadian pulp and paper industry. Pulp and Paper Canada 1989; 90(9):T297-T303

Murdock J.L. and Hayes-Roth G. Intelligent monitoring and control of semi-conductor manufacturing equipment. IEEE Expert 1991; 6(6):19-31

Qian D. An improved method for fault location of chemical plants. Computers Chem. Engng. 1990; 14:41-48

Rao M. and Corbin J. Intelligent operation support system for batch chemical pulping process. Engineering applic. of Artificial Intelligence 1993; 6:357-380

Rao M. Integrated system for intelligent control, Springer-Verlag, Berlin, 1991

Rao M. and Xia Q. Integrated distributed intelligent system for on-line monitoring and control of pulp processes. Canadian Journal of Artificial Intelligence 1994; Winter Issue:5-10

Shafaghi A., Andom P.K. and Lees F.P. Fault tree synthesis based on control loop structure. Chem. Eng. Res. Des. 1984; 62:101-110

Stephanoploulos G. Artificial intelligence in process engineering–current state and future trends. Computers Chem. Engng. 1990; 14:1259-1270

Soucek B. From modules to application-oriented integrated systems. In: Neural and Intelligent Systems Integration: Fifth and Sixth Generation Integrated Reasoning Information Systems. John Wiley & Sons, Inc., 1991, pp 1-36

Xia Q. and Rao M. Fault tolerant control of paper machine headbox. Journal of Process Control 1992; 3:171-178

Xia Q., Farzadeh H., Rao M., Henriksson C., Danielson K. and Olofsson J. Integrated intelligent control system for peroxide bleaching processes. Proc 1993 IEEE Conference on Control Application, Vancouver, Canada, 1993, pp 593-598

Yestresky J. and Ziemacki M. Fuzzy data comparator with neural network postprocessor: A hardware implementation. In: Neural and Intelligent Systems Integration: Fifth and Sixth Generation Integrated Reasoning Information Systems, John Wiley & Sons Inc., 1991, pp 323-332

Yu C. and Lee C. Fault diagnosis based on qualitative/quantitative process knowledge. AIChE J. 1991; 37:617-628

INDEX